I L

O V

E   B

E   V

Through the ages, many men have tried to say that Einstein was wrong. They all failed!

That Einstein's postulates are so widely accepted is obvious in the refusal of science journals' to accept any work which purports to refute Einstein.

Historically, science has always been hampered by the elitist few who strive to censor new discoveries.

But we must ask ourselves one question: If Einstein was so right, why has science, once again, bogged down in a quagmire of redundancy?

In this book, Atkins goes beyond Einstein. Starting with a proclamation that Newton had it wrong, Atkins explains how Einstein could not have been right because dear old Albert constructed his theories based on Newton's theories.

Sound like science fiction? If it is, it is on a level so far removed from current-day thinking that a whole new field will have to be contrived just to contain it. You do not want to miss this one! It is science fact in the making.

Physics books are going to be rewritten...and you can take that to the bank!

UNIFIED FIELD THEORY...
Extreme
Physics

by John Hildreth Atkins

Copyright 2006 by John Hildreth Atkins
ISBN **978-1-84728-439-6**
UNIFIED FIELD THEORY-Extreme Physics

Published by Atkins Enterprises
Salem, Oregon, USA

All rights reserved. No part of this book may be reproduced or transmitted in any form or by any electronic or mechanical means, including photocopying, recording or by any information storage and retrieval system, without the written permission of the Publisher, except where permitted by law.

DIMENSION "O"
(or The Evolution of Thought)

Table of Contents

Part One: The Beginning

Introduction..Page 009
My Stupidity..Page 013
More Stupidity...Page 049
Transcending Stupidity..Page 059

Part Two: A New Beginning

Introduction Revisited...Page 083
Space as a 3-d Entity..Page 089
The Big Bang Booms..Page 095
A Look at E=MC (squared)...Page 101
Einstein Goofed..Page 109
The Argument for Attractio..Page 113
Einstein's Second Mistake...Page 117
Redefining Gravity..Page 119
A Look At Dimensions..Page 127
Black Hole Observations...Page 129
Speed of Light In Curved Space................................Page 133
The Absorption Theory..Page 137
Proving the Ether Exists..Page 141
Escape Velocity...Page 143
The Bigger Picture..Page 147

Part Three: The Future

Unified Field...Page 157
Gasoline Engines Are Obsolete.................................Page 165
Global Warming..Page 167
In Conclusion..Page 170

PART ONE:

In the beginning…

INTRODUCTION

We all have our secrets; things which we really prefer to keep locked away from prying eyes, lest we be viewed in a lesser light. For me, my early ramblings about Unified Field are an embarrassment which I prefer to forget. Nonetheless, I am being objective and am including those peculiarities within the confines of this work so as to preserve it for posterity.

Bear in mind, as you read this book, that my early stuff was the product of my ignorant beliefs coupled with very little knowledge. As I gained in knowledge, hopefully, my intelligence took leaps away from my idiotic babblings.

I, personally, do not see any benefit (to myself) in presenting my "stupid" period. Obviously, there are a lot of things in support of any decision to "coverup" my blunders. But one point did occur to me. A very valid one.

A lot of people out there, too many people, will back away from a project because they feel they are not worthy of the task. They rationalize that there is little point in tackling problems which very educated men could not come to terms with. And that is why I decided to enclose my early rantings.

You still don't get it? Well, you see, if a person as dumb as I, obviously, was, could accomplish such monumental feats, then there remains the last bastions of hope for those who follow. It is my desire, and my sincerest hope, that my honesty induces uneducated people to undertake anything that they are passionate about.

We only fail when we tell ourselves that we are failures. Never give up. Moreover, dare to think differently than your fellow, and don't be afraid to take that giant leap of faith. In the final analysis, winners are not made by accomplishments as much as they are made by struggling.

Every education is wrought with the pratfalls of error. If we never made mistakes then we would suffer the illusion of thinking we are always right. But, by making boo-boos, we learn. And that is the best that any of us can do.

As I was in the process of writing this book, I was sure that a few of my theories were reality. Others, I was not entirely sure of. The odd part was that, once I began to delve into this with a serious effort, the pieces of the puzzle kept falling into place.

One example of a doubtful scenario came with my realization (hunch) that no force of attraction exists as such. I was faced with the obvious effects of the Moon on tides and, harder still, the power exhibited by a pair of magnets. As it turned out, the Moon was the solution...and not the problem.

The problem of the magnets is going to take a bit more finagling but I am confident that they, too, will come around to support my supposition. We shall see.

The nice thing about all of this, at least from your point of view, is that you do not have to waste time on my early postulates. You can fast forward to the good stuff by turning the appropriate number of pages. After that, if you still have half a mind to matriculate the untenable, by all means, turn back to the beginning.

I was going to devote a huge amount of time to doing the math, as well as taking each theory as far as I could, before releasing this book. Instead, I am releasing it in a, rather obvious, incomplete or amateurish fashion. For example, there is no rhyme or reason, no order to the manner in which I present my data/theorums. I left it that way because I was, literally, developing these as I went along.

As a matter of fact, I only completed my Unified Field Theory yesterday morning. As interesting as the theory, itself, is the way in which I progressed towards it. So I leave the manuscript exactly as I first derived at it…in sequence, for the most part. Perhaps this will help some hotshot develop a hypothesis for how genius manifests. And my apologies for being so vain.

Therefore, in the profound interests of both curiosity and science, let us begin our ardent journey into the abyss of indeterminate comprehension.

Part One:

My Early Thoughts:

Realizing that scholars are amongst the minority of my readers, I feel a certain obligation to explain what a Unified Field Theory encompasses. That is to say: What the heck is it?

We are all familiar with Newton's laws of physics. They are the basic laws which we all grew up being vaguely aware of. Sir Isaac Newton's. They pretty much explain the world that we see around us. They are the fundamental laws which say that the faster I go in my car, before slamming it into a tree, the more likely it is that I shall be killed by that collision.

Obviously Newton didn't say that. The man lived in the 1600s. The closest he could have come to a car was a horse and carriage. The point I was trying to get across was that Newton came out with the 'easy' laws about physics. Things like, "for every action there is an equal and opposite reaction." Or, "A body in motion tends to remain in motion (until acted upon by some other force)."

Newton's first law of motion is often referred to as his Law of Inertia. It says that a body at rest tends to remain at rest and a body in motion tends to remain in motion...unless acted upon by another force. The simpler way of stating this is to say that whatever a body is doing, it will continue to do until some force compels it to do otherwise.

Newton's second law is the acceleration of an object, as produced by a net force, is directly proportional to the magnitude of the net force and is in the same direction as the force, and inversely proportional to the mass of the object. "Net Force" is used here to represent the dominant or prevailing force upon an object. This distinction is necessary because there is always more than one force acting upon a given thing at any given moment.

One of the interesting things about Newton's second law of motion is that it implies that the faster a body moves, the more its' mass/force increases. This implication establishes that energy is converted into matter and insinuates that that matter can be converted into energy. As we shall see, shortly, there really is only one distinction between energy and matter.

Newton's third law of motion is probably the one that we all most remember from science class. For every action, there is an equal, but opposite, reaction.

There is an inclination to accept that all forces are symbiotic. They are either push, or they are pull. And one simply does not exist without the other. In this sense, Newton's third law is the most important one for us to remember. You see, the whole Universe exists as a symbiotic (balanced) entity which is comprised of symbiotic (balanced) units.

From Newton we will take a huge jump to Albert Einstein and his theory of Relativity. Basically what Einstein was attempting to expound upon was the laws which govern Quantum, or Mechanical, Physics (subatomic particles such as atoms and quarks), as well as Classical Physics. Einstein was searching for a thing that he called the Unified Field Theory. This theory, which eluded him for all of his life, was supposed to be the Holy Grail of Physics. Why?

The Unified Field Theory was supposed to unite Classical, or Newtonian, Physics with Quantum, or Mechanical, Physics. Albert felt that there should be some formula, or two, which might be used to calculate for every force in the universe (including matter). He failed at his self-imposed quest.

Einstein was a very remarkable fella. To really appreciate what he did for the world of science, you need to understand what he was faced with at that time.

Very few houses had electricity. A simple light bulb was awe-inspiring. Most Scientists were only vaguely aware of such things as atoms, electrons, photons, and even household germs. All in all, the world was a pretty crude place to live. Yet, somehow, Albert managed to look around him and see things as they really were. So don't, for even an instant, ever deny that great man his due.

And never forget that Newton was even more in the "dark" ages than Einstein. Great men have always seen beyond that which was around them.

My world was very different from even Einstein's. I was raised to believe in atoms, nuclear physics, computers, microwaves, etc., etc. These things would have been shocking in Albert's day. And I am not all that sure that we aren't still fascinated by them (on some level).

What I am saying, in my usual long-winded way, is that our world is different from Einstein's and that makes looking at it from fresh vantage points all that much easier. In the following discourses, I have attempted to explain the way I see the Universe in terms that are both acceptable to Physicists and still be understandable to the man who doesn't care all that much for such heavy thinking.

Finally, while I do not, yet, possess the ability, or desire, to express my ideas in numerical or symbolic equations, I feel that anybody with a few years of training ought to be able to do that for me. I guess the proof will be in the pudding. In any event, I hope you find it entertaining and insightful. After all, I am a Writer and that is my job! So let's get with it.

Before anyone can formulate a Unified Field Theory, we have to alter our perception of the atom. Since the atom is the basic unit from which all things evolve, it is necessary to understand it before undertaking something as monumental as Unification. In any event, we are going to utilize the K.I.S.S. Function (acronym for Keep It Simple, Stupid).

To my knowledge, nobody has accurately defined an atom. As I look at the world around me, I see the workings of an irrefutable law. Newton referred to this law by saying that for every action, there is an equal, but opposite, reaction. He may well have carried that concept a step further and simply stated that for every positive, there exists a negative; for every hot, there is a cold. One cannot exist without the other because we live in a balanced Universe where symbiosis is the norm and not the exception.

I am recalcitrant. I could never accept that the Earth is flat anymore than I can accept that the atom is as my worthy predecessors supposed it to be. The naïve manner in which the atom is described in textbooks leaves so many holes which, in turn, leave holes in their ensuing theorems. If we are to have a chance to ascertain a Unified Field Theory, we must, let me repeat that, we MUST, redefine the atom with greater precision and clarity.

There are times when science is not going to be adequate enough to explain a particular phenomenon. What is required in those instances is something that is in much shorter supply. At those times, we all need a heavy dose of something called 'Common Sense.'

Can we, for even a moment, actually believe that the atom, the very foundation of all that exists, is so chaotic and unstable as we have, thus far, been inclined to believe? I think not.

The picture that we should have of the atom should mirror precisely what we see in the Universe around us. Here, we are discussing the basics. Let the Sun represent the center of the atom. With the Sun being symbolic of the nucleus, we can progress to include the planets as symbolizing electrons. We can now begin to visualize the basic structure. Let's analyze it.

The Sun spins or rotates (if you will). The planets do likewise. In fact, everything in the known Universe spins. Is there any reason to presume that the lowly atom acts any different than the Universe itself? Or vice-versa?

The short and simple explanation of this characteristic is that everything is a form of energy and energy equals motion. Nowhere in the (known) Universe is anything purely at rest. So the question begs: Does the atoms spin because the

Universe spins, or does the Universe spin because the atoms spin?

For the sake of making calculations, it really makes no difference. In the real world, one might compare that to which came first; the chicken or the yoke? The answer is the yoke. So, if you made those comparisons, you goofed. You see, the other answer is that both act one upon the other. (K.I.S.S..) A chicken is a finite creature which began as an embryo. The Universe is infinite (no beginning and no end).

Which came first...energy or matter?
Can you even imagine the implications, not to mention the ramifications, involved in the notion that something could actually be in a state of rest? By rest, I mean that it does not move relative to anything else.

If such an event were to occur, it would then become the center of the Universe. It would become relative to all things and all things would become relative to it. Time would freeze.

At the same time, simultaneously, it would have to be impervious to every force known to man. This would be an oxymoron, or paradox, of hitherto unknown proportions. How can matter exist without any interaction whatsoever with the elements surrounding it?

The answer is, it can't. At least, not in the classical sense. In a relativistic mode of speaking, we can say that something is at rest when compared to another thing. That simply means that each is moving in a direction, speed, and illusionary fashion to project the impression that one is at rest relative to the other.

But we need to get beyond the concept of relativity if we are to develop our Unified Field Theory. Let us take a closer look at the building blocks of the Universe.

Irregardless of what we choose to call the various parts of the atom, we do need to be aware of how and why they exist. So let us begin with the nucleus.

Nuclei come in all sizes. Notice that I did not put any restraint on the mass of an atom. I did this with the contemplation that, for recogni-tion of events, we might even consider the solar system to be an atom and examine how it interacts in the grand scheme of things. But that is too far afield at this juncture. Just keep it in mind for later examination.

As I was saying, the nucleus of an atom varies from atom to atom. I do not foresee that any of my esteemed colleagues has any problems with that. So I will continue.

We now know two things about the atom. We know that its nucleus can be any size and we know that it spins. What we do not know is what the charge of the nucleus is, the speed of its rotation, and the direction of the rotation. We can calculate each of these things. And we can do so with a fair amount of accuracy.

But we have one handicap which has been blocking the proper research. I am, of course, referring to the idea that the nucleus of the atom is made up of protons and neutrons and that the nucleus, therefore, is positive. Which leaves us having to believe that the electrons are negative. This is so grievous and outrageous as to render the whole concept preposterous.

Ladies and gentlemen, we cannot make advances in physics if we have, collectively, buried our heads in the sand. Is our Sun the same as every other sun in the galaxy? Is each planet in our solar system the same as the others? You see these things with your own eyes and yet you refuse to believe that they can occur at sub-atomic levels? How can you ever hope to connect Newtonian principles with Quantum physics if you refuse to see the truth?

Universal laws are universal. There is not one law for one thing and another law for another thing (unless you are a poor man going to court against an attorney). What works in the galaxy also works in the sub-atomic world of Quantum physics. What Scientists appear to be doing is trying to rewrite history before it occurs.

When you look at the solar system, and you can see what is going on, and you can't explain that, then you have completely missed the mark in

Quantum physics. There is nothing mysterious about anything that is going on in the solar system. It is just that science, like everything else humans touch, is being portrayed as being more complex than it necessarily is.

What I want to know is, do we make things more complicated because we think things are smarter than we are, or do we make things more complicated because we think we are stupider than those things? Guess it is a matter of perspective and subjectivity.

Let me explain the solar system to you so that you can better get a handle on the atom. We tend to be creatures of ease. We look at the solar system as being a unit of space sporadically inhabited by clumps of dirt. And we have a sun that illuminates the whole shebang. Isn't that pretty much how you look at it?

But what about all of the forces at work in the solar system? Space is fluid. We know this. Our initial mistake is that we forget that it is a river. Like every river, it consists of two parts. We have the water and we have the river bed. We can all see that we have to substitute light for water. But what do we substitute for the river bed?

For now, let's just suppose that the "ether" of space consists of atoms which are comprised only of nucleus's. There are no electrons. This is not, technically, correct, but we will use it for supposition.

Each of these electron-less atoms has a mass which acts upon every other thing with mass. Each mass is a compilation of charged particles. So far, it is all pretty basic. Simple. But this is where most people get tripped up. Define charge?

What is magnetism? Other than saying that it is a phenomenon, scientists have been reluctant to explain magnetism. What is it? Why does it work? Well, I told you that there are no mysteries in our system. Magnetism is simply a direction of motion of a group of atoms, all having the same, or similar, magnitudes. Opposites attract; likes repel.

There is no such thing as a body at rest in the Universe. Therefore, there is no such thing as a neutral, or non-charged (non-moving) particle. The effects of forces canceling each other out, momentarily, gives us the illusion of neutrality. That is not to say that there are not particles out there which are independent.

I believe that our minds are tapped into a body of collective knowledge. I also believe that the lowly atom is also tapped into that body of knowledge. It knows what its' job is. So, when I reference to an "independent" particle, I am referring to one which is not connected to that same body of knowledge.

In review, we know that the nucleus comes in various sizes and each spins. We also know that a nucleus can spin either right or left. And we know from this motion that the difference in direction determines 'charge.' Now let's make it complicated.

An atom can also rotate up, down, or in other dimensions. And they can be doing this simultaneously to left-right. In fact, this up-down spin can be seen as a whole different polarity than simply left-right. And we can add to that forward-backward. Considering only the factors of size and direction(s), we have a rather large number of possibilities as regards the behavior of each atome. But we are not even close to done yet.

Continuing on, we have fast spinning and slow spinning and, probably, a lot of points in between. In theory, we have an almost infinite possibility of combinations. Now let's add the electrons!

The electrons have all of the same properties of the Nucleus with one major exception. An electron is limited in its size by the mass of the nucleus. In ordinary terms, a moon cannot be larger than the ability of the planet's energy to hold it in orbit. We cannot overlook the inverse square relationship on any level. It is the same at the sub-atomic level as it is at the planetary level.

The possible combinations of electrons and nucleus gets mind-boggling in a hurry. Though not infinite, these can be very complex. The good news is that certain combinations have the same effect as other combinations so we, from a causal standpoint, do not have to deal with each individual atom.

A major problem faced by the contemplator of these (ir)rational ideas is the concept of a space which acts to confine a body while, at the same time, transmitting energy from one body to another. This may seem chaotic and fractal, in that all matter has a duality. Here we go again (eek).

Remember that in all of nature, we cannot have negative without positive, hot without cold, north without south, etc., etc.

So if I place two north-seeking magnets in proximate vicinity to one another, they are going to push each other away. If we have them dangling by strings in balanced fashion, the larger, or more powerful, of the two will succeed in spinning the other magnet around. But there is a way that the weaker magnet can win out!

If we were to put the weaker magnet into a fast spin, and we also put it in an elliptical orbit so that its north-seeking pole always presented itself to the other magnet's north-seeking pole at the same precise time that the two were in closest proximity, the accumulative energy of the spinning magnet would be greater than the energy of the larger magnet, and the little magnet would win the contest.

Let us now return to the solar system as a river. In physics, there is a phenomena known as a Flettner motor. Water flowing through, as a river, will exert more force on some parts of the river bed than others. On Earth, we can see this as erosion along river banks which causes the river to meander in a series of twists and turns. Are you with me?

If you think of the river as the energy (light) radiating out from the sun, you can easily deduce that this light energy would impact the planets, as well as every body in the system, in a predictable manner. Planets rotate because this energy exerts greater force on one side of the planet than the other (not to exclude latent energy or other tendencies induced by rotating atoms). Planets are asymmetrical. Light curves for the same reasons as the river bends. Let's take a closer look at light.

Light. 186,000 miles per second. Thought to be the fastest unit of any kind in the known Universe. Also thought to be a steady speed which cannot be slowed down or sped up.

Ridiculous. Preposterous. Humbug. Men are finite beings which sees the Universe in finite terms. Quantum Physicists entertain themselves by conjuring up equations, theories, and the like, in a mesmerizing attempt to explain infinite expressions in finite terms.

Scientists confuse us by stating that light is a string of photons which travel in a straight line; something akin to electricity in a length of wire. Indeed, they postulate that both light and electricity travel at the same rate of speed.

I believe that the whole Universe consists of an ether or, for simplicity, light manifesting as matter.

If, for example, a tree was not light, then a laser could not effect it. This is because a given force can only work on a similar, or opposite, force. I believe that there are super fast "energies" that we cannot detect. The same thing is true of super slow energies. There is a narrow band of frequencies (or dimensions, if you prefer) which we humans can see, hear, touch, smell, or taste.

I do not perceive light as being a string of photons but, rather, as a series of reactions in sequence or chain. These reactions occur at speeds far greater than 186,000 mps and along the path(s) of least resistance (straight line), usually.

If you are thinking (as opposed to rebuking), then it should be obvious that the notion of converting matter to energy, and vice-versa, is more a matter of slowing down or speeding up light. And there is a lot of substantiating evidence with which to support this concept. I.E: Magnetism and, more specifically, electromagnetism.

I once advanced the theory that energy is matter in motion and matter is energy at rest (but not 0). We also know that the faster a given quantity of matter moves, the more heat (energy) there is. Conversely, the slower a given unit travels, the cooler it becomes.

Magnets are generally cool. And they cease to exist if they become too hot. It is also noteworthy that the best examples of both magnetism and electricity involves metal.

I advanced another theory that as the speed of light is slowed down, there is an increase in its mass (force). People like to scoff at that idea but the explanation is relatively simple and not at all complex.

If light consists of a series of photons (as other researchers claim), each following the other, physics tells us that the first one expelled from the source is going to encounter resistance from the ambience. This resistance will cause the first photon to slow down. Each successive photon is still traveling at 186,000 mps and through a hole, or path, swathed by the first photon (path of least resistance). The result is a "pileup" just like the ones you hear about out on California freeways.

Another aspect of this theory is provided by the fact that light is a charged particle. All kinds of neat things happen because of this charge. Light does not travel in a straight line. It serpentines like a snake. This gives it wavelike properties that can really irritate researchers.

I was privy to enter into a discussion with some learned colleagues one night and one of them inquired as to what was this Blarney Stone that would slow down a ray of light? It was a valid question, and one that I had rolled around in my head on several occasions.

Perhaps it was a lens made of layers of two, or more, types of alternating materials (like in the Great Pyramid of Cheops). Maybe it could be stored like current in a capacitor. The consensus of opinion was that there had to be a substance, whether manmade or natural, which would allow the manipulation of light energy.

Absent any definitive substance with which to work with, we looked at other available resources. We envisioned a laser-like device which ran light through a series of electromagnets which were pulsing or resonating in harmony with the light wave. Once synchronized, it would simply be a matter of reprogramming the power source (which would be gigantic) to lower the pulse or frequency of the light.

The idea of setting magnets up at the opposite ends, encircling the mirrors, was debunked as useless. It was thought that the shift in momentum in reflected light would be so imperceptible that electromagnets would have zero affect on it. The solution would not be something as easy as that.

Another idea that crossed the board was that passing light through various mediums, in conjunction with mirrors and magnets, might be the key. While that may seem possible, it doesn't seem plausible. My feeling is that a medium is only good for changing the frequency of the light and does nothing to slow it down. But the medium may prove useful if it is ascertained that one frequency of light is easier to control than another.

For the Novice, or laymen, out there, all light, if uncoiled, is the same length. At least that is the prevailing theory. The thing that distinguishes red light from blue light is frequency. You might think of that as how many coils does it have (just as a snake coils as it serpentines from one place to another). Stretched out, red light is the same length as blue light. So as I speak of slowing down light, I am not talking about any one light (color) over another. I hope that clarifies things.

And for you Physicists, I should reveal that light particles have different sizes and those variances in sizes are what affects frequency. Lower frequency light is made of larger particles and higher frequency light is made of smaller particles. You could have predicted that, if you would have given it a little thought instead of accepting the frequency thing as Gospel.

If, as I suspect, UFOs operate by manipulation of light, then the unit that controls light has to be smaller than, say, the humungous power source required for a powerful laser. Light propulsion would have to rely on pulses. Ironically, or interestingly, this concept comes from the Jetsons (cartoons). And the idea of bottled light is straight out of Star Trek.

It would seem that I fantasize way too much because I have enlisted cartoons and space fantasy as examples for futuristic travel. Not at all. Our minds, more specifically nine-tenths of our minds, are functioning on a higher plane of existence. Those space fantasies and fanciful cartoons were created from data collected in this higher plane. Why is that so redundant?

One of the things that annoys me the most is all this hoopla over the big bang theory. I find that such an utterly preposterous notion. Space is occupied by two essences. There is the so-called ether, which is fluid, and there is light, which consists of particles. Simple

Newtonian Physics tells us that a particle traveling at high velocities has to have a mass force. And that mass force must act upon other masses, propelling them in the same direction of travel as the photons.

Light travels away from the sun and is reflected away from every body in the system. This force pushes every planet away from the sun and every planet away from each other. Moreover, over time, each body must increase its mass and thus, according to classical physics, the increased mass will cause the bodies to increase the force of attraction between them.

However, the sun must be losing mass and that loss in mass would result in the planets moving away from the sun, as well as each other.

With two, seemingly different forces pushing the planets away, we would expect that the combined forces would accelerate this repulsion at a rate that would be substantial in expanding the universe around us. Big bang theory? Humbug.

If you have to have an explanation for the creation of our solar system, galaxy, or whatever, try the following. Picture an ether composed of a negative particle spinning next to a positive particle next to a negative, next to a positive, on and on, ad infinitum. Along comes a large body such as the Sun and cuts a path through these spinning particles. What is the result?

If we assume that nothing was spinning before the arrival of the mass, meaning neither the particles (atoms) nor the Sun, there would have been no choice but for both the sun and the particles to start spinning (frame dragging). This is highly probable.

The reason that I suspect that the atoms were already spinning is two-fold. Atoms are energy and energy must move. For practicality, the atoms were sitting and spinning, and not randomly shooting here and there.

The other reason that I suspect the atoms were already spinning is because their cumulative spin energies, the very energy that was locking them into an alternating charge state, would have been sufficient enough to slow the sun and hold it in position.

The speed limit of space, for our purposes, is always going to be 186,000 miles per second in a direction away from the sun. The trap that most people seem to be falling into is that they think of light as a force rather than a reactive force. What I mean by this is that a force is something akin to a rock traversing through space with the ability to cause damage to another body it encounters.

Conversely, a reactive force is one that acts as a chain of magnetic responses which do not collide in the conventional sense.

Think of light as being a string of bar magnets which are all connected in a fashion similar to running water. Whereas the water has cohesion properties which have yet to be defined, light is not so mysterious.

The first magnet out lines up with the positive pole to the right. This magnet has a magnitude of 7 (on a scale where 10 is optimum). Next comes a magnet which has a magnitude of 2 and has the positive pole to the left. Then comes a magnet where the positive pole is to the right and the magnitude is 6. Then another magnitude 2 left, 6 right, 2left, etc. etc. Each succeeding magnet flip flops in like manner ad infinitum.

The disparity between the first (magnitude 7) and all of the trailing magnets is due to the initial release being greater. That is to say, it took a magnitude 7 to escape from the host. Think of it as air trapped inside of a balloon and we keep adding air until it bursts. The burst is more powerful than at any time afterward.

That phenomenon which the distinguished thinker, Albert Einstein, referred to as an "interval" of space, is not really space at all. Think of it more as a weak link. And the bar magnets are not bar shaped but spherical.

At the instant that the first photon escapes from the sun, its tension releases a burst of acceleration into fluid space. It meets a bit of resistance and slows down. We view this as the mean speed of light.

Many scientists speculate that if we were to follow a beam of light out to the furtherest reaches of our solar system, we would see that it slows down even further as it spreads out and dissipates. I am, partially, inclined to agree.

People will, naturally, argue that I have shot a hole in my own theory. That is to say, if I assert that the mass force of light increases as it slows down, then wouldn't the slow light exiting the solar system cause grave harm to other bodies it encounters? No, not at all. The reason is simple. The mass of the light is absorbed by the ambient. And this ambient includes other light rays.

Now here's the mind-blower. Light, or certain light (such as that leaving our solar system), will accelerate to something just below infinity. It does this of necessity. Light traveling to us obtains a maximum speed of 186,282mps. If light surpassed that in any appreciable manner, light would pass us by as though we did not even exist.

Since each part of our Universe needs to maintain "communication," light increases its speed to cover the great expanses between stars. Upon arriving at its preset destination, it interacts with existing forces and reduces speed.

Can we prove that the mass force of light increases with a reduction in acceleration? Yes. We have the technology. We can pulse the light at any frequency with the aid of computers. We could, in theory, use coils of lights (as opposed to strong magnets) to control light emission. When Captain Kirk tells Scotty to engage impulse engines, perhaps he means to switch pulsing frequencies.

In the simplest of terms, I believe that all energy/matter originates as light. Remember, I asserted that as light (energy) slowed down, its' mass/force increased. In other words, it began to revert into matter.

Because it is so pervasive, light behaves as a wave. Think of the solar system as being a giant tube and light as water running through it. We know about Flettner engines and the role that flowing water has on stationary bodies. All planets, etc. appear stationary to an object traveling at 186,000 miles per second. I suspect that it is this (Flettner) phenomenon that helps make the planets rotate.

Another factor that appears to be at work is simply the nature of the medium. Different objects, whether due to mass density or the materials of which they are constructed, react differently to the water (light) flow. There is nothing so mysterious about any of this. Anyone who has seen a radiometer at work is aware that a very small amount of light can move physical matter in a vacuum. So why can't it do it in space?

My next thought is that gravity is a consequence of light. We can measure the forces at work and factor those in with light energy and compute gravity. This is immutable. Think of it as light being lines of force (duh) through space, and the planets, etc., cut through these lines of force creating what?

Electromagnetism is not such a hard bird to fry. (Gravity is a weak force because it is not as concentrated as in a magnet and it is acting upon (dis)similar forces).

What *is* hard to cook with is the revelation that the sun is a ball of ice. I catch all kinds of hell over this one and yet it is one of the most obvious things to me. Even absent the theory that hot cannot exist without cold, it just plain makes sense.

Conventional physics tells us that fire must have air in order to burn. And quantum physics allows for critical mass. But, in either case, there still remains one obvious feature of the incendiary device. Namely, the mass of the sun does not shrink in any appreciable manner.

Ice is a solid. It can melt off as a liquid and burn as a gas without any discernable alteration in its dimensional properties. Of course, all of this is contingent on what the material is that makes up the Sun's core. And are there any impurities in it?

Hey, I never claimed to have all of the answers; just the hard ones! Let somebody else do the mop up.

Scientists and their theories about red dwarfs, white dwarfs, blue dwarfs, and the seven other dwarfs, are going to have to revamp their thinking. As these balls of ice burn down and the material at their cores becomes less dense, there is a tendency for these balls to explode. One final flare-up and, poof, they're gone!

Their color is not an indication of thermal intensity as much as genetic makeup. And, considering the immense distances that this light has traveled, can we even accept that it is not tainted upon arrival?

I should like to return to the atom for a moment. Chemists expound on the funda-mental concept of covalent, and/or ionic, bonding and hawk the belief that an excited atom has electrons which rise to upper levels and then fall back down again. I believe in those things occurring.

Spinning electrons can bond with two or more atoms. But the thought that we tease an atom and that forces its electrons to go to upper levels is a bit much.

An atom can be influenced by a force, or forces, which cause it to rotate at a faster rate. We know that centrifugal force will force an electron to fly away from an atom. So it is only logical that an 'excited' atom is merely one in which it spins so fast that it casts off, or tries to cast off, electrons.

I wanted to clarify that because there is a tendency to suppose that an event occurs where the atom is heated and this heat is what forces the electrons into more distant orbits. While I do feel that thermodynamics is one of the keys to the secrets of the universe, I do not subscribe to that concept in defining what takes place within an excited atom.

We know that atoms which add, or lose, electrons can mutate into other elements. It is this fractal logic which compels one to believe that over the course of millions of light years, there is a strong possibility that the light reaching us from distant galaxies may have suffered a mutation or two. In fact, I would count on it.

If a scientist, or physicist if you will, should look at an event and ignore even one possibility, he has failed to do an adequate research of the phenomena. The omission of a single clue is tantamount to trying to solve a riddle with every other word missing. Possible, but not likely.

Let's journey back to the fabric of space or the so-called ether. I compared it to a river bed and stated that it was comprised of atoms minus their satellites. This would almost seem fourth dimensional in that it appears to exclude the three classical states of matter. It is not a gas. Nor is it a liquid. And it certainly isn't solid. So is it that quantity called Space?

That brings up a most interesting argument. There are those who would claim that it was, indeed, time. Moreover, they would assert that it is intricately joined with space. More fodder for the masses (excuse the pun).

Time is motion. Physicists claim that the fabric of space is stationary and does not move. Thence, how can one claim that it is time and, in the same breath state that it does not move?

There is absolutely nothing fourth dimensional about the fabric of space. It is not in one of the classical states of matter because the influences have not been applied yet. Space can be gaseous, liquid, or solid as a rock.

This is comparative to saying that an embryo is not human because it has not formed a particular sex yet. Is it any less human because it is neither male nor female? Can we not say that it is both male and female until it decides which to be? And why is it we cannot say that space is liquid, gas, and solid (for the very same reasons)?

If you cannot grasp the concepts of the atom and the material of space, then you are going to have a very difficult time understanding my next theory. As far as I know, this one is all mine.

Light does not travel in empty space. Light travels from body to body to body for as long as there is enough energy, and a path, for it to do so. In its simplest expression, light (C) travels to body (a) until (a) is saturated. Afterward, Light will travel to body b, c, d, e, and beyond, for as long as it is able to saturate each and move on.

Now, thanks to Newton's laws, if we have a Sun which expends X amount of energy, or force, in a direction away from itself, there must be an equal force going to the sun. In the current accepted theory of the atom, we would say that the Neutrons carried the protons and electrons to where they were needed and then returned to the sun for reloading. In my model of the atom, there are only protons and electrons and the closest thing to a neutron would be two electrons coupled to a proton. Why?

The proton is roughly twice the size of an electron. It takes two, or more, electrons to balance out a proton. So what is happening is that the light exits the Sun looking for a body in space. It, through a series of complicated calculations, knows which body closest to it, needs cargo. It races there, bangs head-on into whatever mass is present and 'loses' its cargo. The now lonely Proton must seek out an electron for company. So it goes to the nearest one. Think of this as being the path of least resistance. Well, where do you suppose that the nearest electron is?

You guessed it. The nearest electron is directly to the rear of the proton. So it turns around and heads right back towards the sun (or source, if you will). Since the return is done at precisely the same rate of speed, and direction, there exists a certain state of equilibrium in space. Well, almost. But the difference is so negligible as to render it nearly incomprehensible to man.

A question that comes to mind is how can light turn around without losing speed? The answer lies in the rate of spin of the nucleus. Remember my earlier example of the little bar magnet verses the larger, more powerful one? Use that concept to visualize the powerful attraction of one light quanta to another. For all of our purposes, not to mention perceptions, the switch is instantaneous. And that is substantiated by the two slits in the dual aspect experiment of light.

Scientists can calculate the attraction force and use that to calculate the amount of spin for light atoms by using the parallel slits experiment. Anyway, they can do so now that they know what they are looking for. And, if they do not have the vaguest idea of what I am talking about, I will be glad to explain it in detail for any researcher.

I have heard it said that many scientists believe that the Universe or, at least, the solar system 'breathes.' Hopefully, they have an explanation that which they can work with. And, in case they don't, let them ponder this: How is it that light from a flashlight fans out and disappears after only a few feet, but light from a star, millions of light years away, shows up virtually unscathed?

The answer, of course, is that the flashlight was so wimpy that the light was gobbled up by the dark (matter that has a deficiency of energy). The light from the distant star did not fan out into space as there were not bodies enough to attract the light. So the light went from planet to planet to comet, to planet, or whatever. Kind of like a telegraph. The fact that this light gets to the Earth at all, should be enough to prove my theory.

Moving right along, we next come to one of my personal favorites, how come Saturn and the other large planets have rings? That one is so simple that a man might wonder if he lost his mind when he realizes it.

We have a bar magnet. We observe that a magnetic field flows from the North-seeking pole to a point about halfway down the bar. We observe that a magnetic field flows from the south-seeking pole and it, also, extends to a point about midway along the bar magnet. This creates a 'valley' along the equator of the bar magnet. Okay?

Now, the large planets have both these poles and the fields they generate. There is, however, one major difference. These huge monoliths are rotating at fairly high speeds. This means that any (charged?) material that comes into contact with the magnetic fields is going to be pushed towards the valley where the force is weakest. Moreover, since the magnetic force is weak at the equator, these particles are being thrown out into space. So long as the planets continue spinning at sufficiently high rates of speed, those particles will remain as rings.

That theory really makes a lot of sense. However, it is incomplete. You see, we must take into account every force that is acting upon a particle and not just look at the singularity. When a body in space rotates, space rotates. Scientists refer to this as frame dragging.

As the body rotates, its equator is spinning fastest. Its axis tips spin slower. What happens is that the space around the equator is compacted more than the slow spinning space around it. This compaction causes space mass to increase. This increase in space mass enables space to attract and retain matter.

Another way to look at the phenomenon of compacted space is to think in turns of dissimilar speed. The slower space acts as a barrier which keeps mass confined within the parameters of compacted space. Think of it as a wall on each side of the fast moving space.

We see this force at work in the solar system. Each planet is within that narrow band created by the Sun's rapid equatorial rotation. Likewise, we see a planet's moons are within a narrow equatorial band.

I think one thing that confuses scientists is the notion of magnetic attraction. For instance, it is believed that this force of attraction is what keeps our moon orbiting us with the same face towards us.

The reality is that the moon, like everything else in the known universe, is spinning on its axis also. Its rotation is guided by the motion of the compacted space of the Earth's own rotation. Think of that as being similar to a marble rolling on a conveyor belt in such a fashion that the marble can roll with the belt but must remain stationary to everything else in the room. In other words, frame dragging rolls the moon at a pace which causes it to always present the same face to us.

So, you see, that this was not as complicated as you may have supposed. And that is one of the beautiful things about nature. It tries not to be complicated. It tries to take the path of least resistance. It tries to draw you pictures of what is going on around you. All you need to do is take the time to look, think, and take the easy way out. Be like nature.

Now, having already given you the concept that light cannot travel unless it has a source and a destination, and the destination has to be anything other than space, we might want to examine a black hole.

Nothing in modern times has elicited more awe and fascination than the notion that there is this gigantic black hole in space that seems to gobble up energy and matter in huge amounts. In one of the most accepted theories, a black hole has been described as the birth canal of a Sun (or other cosmic body). And, I suppose, that makes a fair modicum of sense for those with such an inkling; particularly, those who have a bent for fantasy and science fiction.

So what is the truth? A black hole is nothing more than a point in time-space which has not been intersected by any light-energy. A simple analogy would be to draw five dots on a piece of paper and draw lines which connect each one to each of the others. Let the dots represent the stars and the lines represent rays of light. What do you see in the places where there are no lines or dots? Space.

If there is no reason for light to be there, why would there be any light there? Nature is not in the habit of doing things for which there is no reason to do. If there is no need for C at x, y, z, z', then nature does not put it there. And man should stop trying to out-guess the old broad.

In short, the Universe is a very orderly thing. Everything has an assignment and an assigned frequency. There are no mysteries; only misconceptions. And if Physicists want to waste time trying to speculate about what goes on in other dimensions (frequencies outside our range of perception), let them. A more erstwhile pursuit would be to analyze the things which they feel hangs in both our world and the other to see if there isn't some way that we can use those things to better control our own Universe.

So, in conclusion, if you wish to formulate the Unified Field Theory, you must not only take in the forces that act within this small spectrum of forces in which we function, but also consider the influence of the forces which border the cusps between the lower range of frequencies and the higher range of frequencies of all that we are allowed to perceive. You might think about it in generalized terms such as E (energy of the highest magnitude) equals M (energy at the lowest magnitude) times C (the constant speed of light at the given co-ordinate) squared. You can manipulate from there.

The important thing to remember is that everything is relative according to motion (speed). The faster a thing travels, the more energy it becomes. The slower it travels, the more mass it becomes.

In terms of the Solar System, that which we call 'space' is not so much an interval as it is a transition point between energy and matter. Scientists speculate that celestial bodies are acting upon one another 'at a distance.' They have devoted a lot of time to trying to define how this is so.

There is a tendency for knowledgeable people to speculate that 186,000 miles per second is the speed limit of the action (moment that light is emitted) until reaction (the maximum velocity of light). This is greatly erroneous. Whatever the event is, it acts instantaneously with the whole universe. Light, for all intents and purposes, is simply the aftershock.

Something to think about.

More Early Thoughts:

The one concept which I cannot stress enough is the notion that everything is symbiotic. And do not forget that it is spinning. Let me give you a simple example of this relationship and the magnitude of its influence.

Scientists speculate that the Earth's magnetic poles reverse about every ten thousand years or so. Not only does that account for the Earth's 'wobble', it reveals some pretty important ramifications. And I am not sure that I am all that versed in them, myself. But let's take a look anyway.

Hold a ballpoint pen in your fingers and start turning it counterclockwise (looking down at it from the top). Keep turning it the same direction as you turn it upside down. Now what do you see?

If you did it right, you see that the pen is now turning clockwise. Did you do it right?

So what does that mean? I can't believe that you asked that, but let's look at the effect.

In my symbiotic universe, everything spins and the difference between 'North' and 'South' is the direction of spin. Yet each is attached to the other so that the opposite for each is directly behind it. Look down the bar magnet and see if that isn't so.

When the Earth's poles reverse, the direction that the Earth is spinning is also going to have to reverse. Instead of the sun setting in the west, it will set in the east.

It is my belief that there were very intelligent people on this planet centuries ago. You can call them Atlanteans, Lemurans, or whatever suits you. Nonetheless, the odds are that they existed. In fact, according to Cayce, they even devised a lens for a ray gun (of sorts) which Cayce alleged that the Atlanteans used to blow up Phaeton (the planet that used to be between Mars and Jupiter).

We know, from a mathematical standpoint, that a planet is supposed to be where the Asteroid belt currently orbits the sun. Each planet is between one and a half and two times the distance of each preceding planet. The Asteroid Belt is thusly positioned. So this would tend to validate the Cayce claim that Phaeton existed at one time.

I believe that the lens for this ray used to sit atop the great pyramid in Cairo, Egypt. And the ancients used the Angular momentum of the energy from the sun to accomplish their feat. The whole construction, and alignment, of the pyramid, reminds me of a huge capacitor. Even the angles of the walls and the materials it was constructed of gives credence to this supposition.

Irregardless of the purpose of the pyramid, I think it important to suppose that there were advanced civilizations on Earth way back then.

And we might even glean some useful information from that hypothesis. For instance, the Bible tells us that God sent word to Noah to build a huge Ark so as to save the various creatures from the deluge of the great flood. Science, herself, supports the belief that a great flood actually occurred.

If this civilization was as advanced as I suspect it was, then it is entirely within the realm of possibility that they were keenly aware of the reversal of the Earth's poles. If that were the case, then it would have been conceivable that they knew that a great flood would take place as a result of that occurrence.

Conjecture? Speculation? Supposition? Sure. But well documented and substantiated by history; not to mention science. And what significance does all of this have on modern man?

Scientists tell us that they think we are nearing that ten thousand year period when the poles are expected to reverse. We know there is going to be a flood and rain which will, if history repeats itself, last for roughly forty days. Anybody got a hammer and some nails?

And what about Jupiter? Here is a planet more than 1300 times the size of our puny little planet. It is revolving once every 9 hours and change. At some point in time, it has to slow down (doesn't it?). If it slows down, won't that cause some, if not all, of its moons to plummet to its surface? And, if that were to happen, would that increase Jupiter's mass to critical? By critical, I mean that it will give the planet sufficient mass to draw in other mass until it, finally, attains enough mass to ignite in similar fashion to our sun.

FYI, if Jupiter were to slow down, the moons would move away. Remember my earlier discussion about a rapidly rotating magnet increasing its force? Same thing. Larger scale.

It seems that, no matter how one starts the discussion, that light is invariably factored into the concept of a Unified Field Theory. As one of, if not the only, driving force(s) in the cosmos, Light is a little hard to ignore.

And why is it that we are always so fascinated with what the fastest object in the Universe is? Does anybody even know what the slowest is? Am I the only one who cares?

As I write this, I am a homeless person who sleeps beneath the stars. Every night I lay there with my feet at the Big Dipper and I ask millions of questions. There is one question for each of the stars above my head. And it is always the same question: How in the hell did light travel millions of light years away from a star and arrive, fully intact, at planet Earth?

By classical physics, light cannot travel great distances without dissipating into the ether or ambience. But we see it with our own eyes. And we cannot explain even the illusion with Newtonian Physics. So what are we to do?

Quantum mechanics is of no use to us as it offers no viable solution, either. Which brings me back to my earlier revelation that light travels in a preset path, from source to destination, without regard for space.

Common sense tells us that this explains why light can travel such tremendous distances without being gobbled up by space. I have, however, proof that this is so.

As I said, I sleep beneath a carpet of stars every night. That is how I stumbled upon an amazing discovery. As a (manmade) satellite approaches a point in space where starlight occupies that space, we can see the path that the starlight takes. We see it because the satellite is unable to influence the light and so it appears to detour around the point of light from the star. Go check it out! It does not happen in every instance but, sooner or later, you will witness this phenomenon.

Even if we did not know about the path being

there, we have to ask why is it that the spacecraft cannot block out the star's light? The answer comes from the other part of my theory. Light will only go where there is a need for it to go. I referred to this as going to a place until it is saturated with light-energy and then moving on to another planet or whatever. Okay?

Now, when the light from the star approaches the spacecraft, it discovers that the spacecraft is already saturated with (star)light-energy (from the sun). Since there is no reason for it to go to the craft, it doesn't. And, since it doesn't, the lesser of the two lights, shifts because of angular momentum (they are perpendicular to each other), and there are other mass considerations. We plainly see the craft go around the path of the starlight.

Scientists have long postulated that light travels in bundles. Indeed, I can see no other way for it travel. If you consider that light is made up of quanta called Photons, with each being something akin to an elongated dust particle, then it stands that the force which propels it away from the source must act upon the largest photon first. The reason I say this is because of inertia.

If we have two particles, a large one and a small one, and they are both traveling at the same rate of speed, and we add a unit of energy to each, which is more likely to break free of the bonds which hold them to the source? It is my argument that the addition of the energy quanta is sufficient enough to increase the mass force to enable the larger particle to break free.

Students of classical Physics will argue that because the one particle is larger, it would take more than one energy quanta to move it away from the source. Under this premise, they will argue that the smaller particle is the first to be propelled away from the source.

It is an interesting argument as stands. However, you must take into account that light is a charged entity. The smaller particles are attracted to, and/or attached to, the larger ones. The smaller particles not only have to fight to break away from the source, they have to fight to break away from the attraction of the larger particles.

The simplest analogy that I can make is to equate the smaller particles to being the rubber band which attaches the larger particles (ball) to the paddle (sun or source). The large particle escapes when the energy becomes greater than the rubber band's ability to hold it in place. This is the classical approach.

In Quantum Physics, the reality is that each photon is comparable to a single cell in human blood. And the Quarks, if that is what you choose to call them, would be the DNA.

Light behaves as a wave because each Photon is told to reunite with other Photons. Imagine that. Light has a brain!

We are able to see stars because it only takes a single Photon to tell us that information. That Photon shows us where it came from. Perhaps I should say, it shows us the last place that it came from. More importantly, it knew it was coming here (before it even left home).

In the dual slit experiment, the light shifts because it knows that it is supposed to be (re)united. Einstein theorized that as an object approached the speed of light, its mass would approach infinity. If we applied that to a Photon, its mass approaches infinity and it is strongly attracted to another Photon because that Photon's mass has also approached infinity. Nice try Albert!

Compared to its size, the Photon's force of attraction may approach infinity, but its mass certainly does not. If it's mass were to reach infinity it would flatten us. The simple fact that we can stand in daylight proves that the Photon's mass is not infinite. So how does science explain this discrepancy?

Scientists have long held that the lowly Photon does not have any mass. Only recently, have they regained their senses and admitted that they were wrong. The Photon does, indeed, have mass. Duh! So, if the Photon has mass, and it travels at the rate of 186,000 miles per second, why isn't its force infinite?

The short answer is because Einstein only had half an equation. Matter and energy are symbiotic.

If we accelerate Matter to the speed of light, it converts to energy. When we decelerate energy, it reverts back to matter.

As the light passes through the two slits, it slows down along the walls of the slit because those walls become Event Horizons as the force of attraction increases. There is nothing so mysterious about it. Just plain old classical Physics. Almost.

There is one little detail that I failed to include. Push. One little ol' word: "Push." Not Pull. Pull is an illusion that can only be defined in our minds. It holds no position in experience. The real world is devoid of the quality we describe as Pull.

Everything in the known Universe is propelled by only one force. It is that force which we have defined as Push (P).

What role does asymmetry have in the functionality of that force "P" as it relates to gravity and other seemingly "pull" functions? That is to say, since everything is asymmetrical, there is an uneven distribution of force P which could account for a realignment of polarities (sic) or other factors which might describe how force P acts as pull.

Can force P exist independent of Time (t)? Time is a measure of motion. If there were no motion, time could not exist. P, in a Quantum or subatomic expression, is motion. Even in a closed co-ordinate system which appears to be in equilibrium, P exists on some level and in some fashion. Reality is that nothing exists in our Universe without P. Moreover, P cannot exist without t.

Even in an open-ended co-ordinate system or other dimensions, P/t exists. P always has three dimensional qualities with a minimum of six possible directions. Left-right, up-down, forward-backward. Or you can substitute in-out for one pair.

Can P exist simultaneously in two different co-ordinate systems? Of course. Even if P were not actually present in one, or both, of the systems, its influence will be. The reason for this is that everything is interconnected.

Scientists have long speculated that we really do not know the outcome of any given experiment because we do not know how much we have contributed to that experiment. Perhaps, just by merely observing an event, we have influenced the outcome. I.E.: to actually "see" a singularity, or other event, we need for light to be present. The addition(al) light may cause the event to change. We simply do not know in which way or by what magnitude.

More on this later.

A Step Backwards:

I would like to say that I have all of the answers. I would like to claim that my work is on a par with the illustrious Mr. Einstein. Mostly, I would just like to know the truth.

Rare is the day that passes without my contemplating the nature of light. For me, it is the cornerstone of the universe. And, while I do have a considerable understanding of its nature, there are questions which cause me great consternation. For instance, what is the relationship of light and fire? Is light fire?

The ancients believed in four elements. Earth, Water, Wind, and Fire. If light and fire are the same entity, then we must conclude that the four elements were Earth, Water, Wind, and Light. And these would be an accurate representation of forces at work. However, they fall short of the mark when it comes to defining the Universe.

We can never forget that the Universe is symbiotic. That is to say, we must be ever mindful that everything that exists has an opposite. Utilizing that supposition, we have to wonder which of the four forces pairs up. What is opposite of earth, water, wind, and fire? If there are only four forces in the Universe, then there must be two sets of pairs.

In a far-reaching sense, we can say that earth and air (wind) are opposites. One is a solid which is, for the most part, at rest. The other appears to be a gas which, again for the most part, is in motion. So, does that mean that water and fire (or light) are opposites? Obviously.

Those who would cling to the ancient belief of the four elements quickly find themselves in disarray when it comes time to ponder the relationship of these to each other. Life may seem simpler if we choose to live with the belief in only four elements. But that is illusionary when it comes to modern concepts and revelations about the world around us. Or is it?

When I look at the four elements of the ancients, I see our universe in a nutshell. If we substitute light for fire, we have a very logical progression. Light is the fastest known entity in the Universe. With light comes motion and with motion comes wind. So we have light, followed by wind. No problem there.

Next comes water. Water is a liquid and liquid flows fairly freely. So we have light (fast), wind (slower), and liquid (slower still). It stands to reason that the only thing missing is "not moving."

I think we can all see that the progression from light to earth is a very logical one. There is nothing so mysterious about it; is there? Well, actually, there is.

The inference derived from looking at the progression of the preceding paragraphs is that there exists some element which can change from Light into Earth. We merely converted the four elements of the ancients into speed zones.

If we take an element "X" and we move it at the speed of light, it is light. If we move "X" at 800 miles per hour, it becomes wind. At 100 miles per hour, it becomes liquid. And when it comes to (near) rest, it is Earth.

There are a trillion souls out there who are, undoubtedly screaming that I have gone off the deep end with my way of thinking.

Perhaps. Perhaps not. I reckon that the best way to sustain my argument is in reversing the process. Sometimes things are clearer when viewed from the other end of the field.

Let's say that I have a clump of dirt in my hand. Let us also say that it is a hot summer day and the soil in my grasp is dry and sparsely compacted. Now let me blow on it. What happened?

Pressure from my breath pushed some of the finer granules and they became air-born. The really fine particles became wind. And if they were to move fast enough, they would become light. This is the reality and not fantasy.

My dear Mr. Einstein postulated that as we increase the speed of a body of matter, we increase its mass towards infinity or, in the classical sense, the amount of force needed to act against that matter would become infinite. Rubbish. I have never heard anything so preposterous and outlandish in all of my life.

Common sense will tell you that as you increase the speed of an object, you accelerate it. This acceleration increases the object's kinetic energy by virtue of the fact that it retains a portion of the energy of the accelerant. So we say that the mass/force has increased due to the increase in velocity. However, we also know that the faster an object goes, the more resistance it encounters.

For some odd reason, Einstein, and associates, prefer to ignore "friction." Even in a vacuum, there must be some friction. Newtonian Law says that there is an equal and opposite force. If we push on something, including space, it must push back. The faster an object travels, the greater the effect we call mass/force becomes...up to a point. After reaching that point (speed), friction or other forces causes the mass of the object to elongate in a direction opposite to that of the mass's direction of travel.

Like Comets heading towards the Sun, an object develops a tail which is comprised of its own mass. From that point (speed) on, the tail gets longer with any increase in speed.

At some point in time, t', the tail gets long enough that it starts to zigzag as it trails behind its host. At t' +1, the tail becomes so long and so thin that it is no longer capable of remaining attached to the host and it flutters away. Now, remembering our earlier discussion about Newton's Law which states that whatever a body is doing, it will continue to do until acted upon by another force(s), we have an explanation for that phenomenon we call "alternating."

Continuing our examination of our accelerating mass, we soon note that it has gotten smaller and smaller with the passing of time (t) and distance (d). I use this symbolism in deference to those who wish to distinguish between time and distance.

The mass continues to shrink until it has reached its lowest recognizable mass-point. It has also reached its maximum velocity or equilibrium. The Matter, for all intents and purposes has now become Energy or, simply, Light.

We accept light as being the speed limit of our universe. That is not the same thing as saying that nothing can travel faster than the speed of light. To my recollection, I do not believe that Einstein ever said that it was impossible for a thing to travel faster than the speed of light. I am supposing that such statements probably came from other theorists, or students, ex-post facto. And Einstein may have kept quiet just so as to appease others.

We all accept the notion that the speed of light is the maximum speed in this universe. We do this because we are limited in our perceptual abilities. And that is fine. Whatever works for us.

But I believe that such half-baked ideas as infinite mass at 186,000 miles per second is subject to confinement in a loony bin. Common sense tells us that if this were so then light would flatten each of us and nothing could exist under the rule of light. In short, light travels at 186,232 miles per second and it does not have infinite mass. And a fellow holding up a mirror does not get pounded to the turf when he lets the light shine on it.

The reality is that the faster an object travels (above a certain barrier which I estimate to be around 160,230 mps), the less mass it has. Common sense will tell you this because we witness it every day. Light has so little mass that a feather stands perfectly capable of stopping it. Just an idea. Something to ponder.

When I first heard of the principle known as the Inverse Square Law, I speculated that there must be a Law which exists at the opposite end of the spectrum. We might refer to this law as the Exponential Square Law.

Light travels in repetitious fashion. We observe lightning striking the ground, retracing its path, and then striking again.

Now picture light doing the same kind of journey. It travels a short distance from the Sun, retraces its step, then journeys a little further out, retraces its steps, then…ad infinitum.

Why would light do this cycle action?

Think of a man building a bridge. He starts from one shore and builds out away from shore. As he leaves material in the bridge, he must return to shore for more material. And so he goes. Back and forth.

The light builds a bridge into space. This is the road it must travel. Let's look at it closer.

Let us say that the light shoves a photon ahead of it. For whatever reason, space takes the photon away. The light cannot travel with a photon and must return to the Sun to retrieve another. This one it carries over the top of the first photon being held hostage. But now, empty space, the second photon is held captive. On and on.

The thing with light is that it does this bridge building with truckloads of photons. And electricity does the same thing with electrons. And this is the foundation from which gravity acts. And all of this has a speed limit of 186,232 mps.

Now light builds this bridge until the bridge is long enough for light to transverse the 186,232 mps barrier. In other words, as the distance light must travel increases, its speed increases.

Many of you have read this book, or more specifically, my discourses on Physics, to see if I can really provide you with a Unified Field Theory. Perhaps I can. More than likely, I cannot. But, in any event, I can give you the direction which will allow you to formulate it.

It is my belief that, by explaining gravity to you, I can get you on the right track. Do you not agree?

Gravity is light. Light has so little mass that it pushes against us and we tend to yawn. It is not the light pushing down on us which causes gravity. It is the absence of light.

Take the case of a comet streaking through the heavens towards the Sun. Does it not have a tail? We, of course, assume that the tail is there because of "solar wind" or whatever. But I submit that the main cause of the tail is the vacuum created by the absence of light.

One of the toughest components of my theories is the notion that there is only one force at work in the known universe. After touting symbiotic relationships, I surprise you by doing an apparent about-face.

The force "Push" creates vacuums which imitate the concept we know as "pull." As long as latent energy exists in the form of matter, Push is always going to exist. Pull, however, is transitory and, relatively speaking, temporary.

When you take all of this into account, and you

reexamine the universe around us, you quickly realize that everything is contingent of field theory. Everything.

I have often thought about how scientists say that the Earth's poles reverse every ten thousand years or so. It occurred to me that logic dictates that something must change in order to set off such a catastrophic event. The only thing that I could think of was that the Earth's mass increases. The Earth is bombarded daily with meteorites and other stuff which increases its mass.

So now I need to figure out how mass affects magnetism and vice-versa. That should be easy as Magnetism is energy and the mass is matter.

But my brain was not finished tormenting me. Now, I have gone on to another theory which I call electron switching. Until recently, theorists in the laboratories had postulated that an atom consists of protons, electrons, and neutrons. They theorized that the nucleus contained one neutron and one proton (coupled) and was, thus, positively charged. The orbiting electrons were, naturally, negatively charged. Well, they finally figured out that this is not true.

But these learned gentlemen still insist that the center of the atom is made up of protons and neutrons...but some protons are negatively charged and some electrons are positively charged. That is preposterous. Utterly nonsensical.

Because it moves faster, an electron is, indeed,

smaller than a proton. I have a very good theory which explains this, but I won't bore you with that at this time. However, in my latest theory, the proton and the electron(s) switch out. That is to say, they trade places. And, like the Earth's poles switching, this is contingent upon parameters defined by the mass of the host.

One of the most fascinating things in physics is the relationship between electricity and magnetism. It is fascinating to me because people just don't get it. We list each in a separate category when the reality is that they are one and the same entity. One is simply slower than the other. This is the simple version. The reality is a bit more complicated.

In nuclear physics, a nuclear reaction is contingent on a lot of factors. One of the most influential factors, other than mass, is 'slow' neutrons. I think pretty much everybody in the civilized world has heard about the importance of the slow neutron. Fast neutrons were not able to sustain a nuclear reaction. We needed slow ones.

An example of this would be if we take give somebody a heavy object like, for instance, a brick. Now we tell that person to stand out beside the highway and we drive by at one hundred miles per hour. Our objective is to snatch the brick from that person's hands. Well, obviously, at one hundred miles per hour, we are not going to be able to do this task.

Now let's go back and drive by the person at one mile per hour. We can easily reach out and take the brick from his hands. No problemo.

As you go through life, remember this simple analogy as it is important on all levels. It is one of the key essentials to formulating the Unified Field Theory. Very important. All things interact favorably with things going in a similar direction at a similar speed. But as soon as you begin to change either speed or direction, watch out.

One of my earlier theories was that all matter is a form of light. We have slow light and we have slower light. The slower the light, the harder the substance.

We take a piece of steel. It is cold (indicating that light/energy within it is moving very slow). As we heat that steel up, it begins to generate electricity. Both electricity and heat are forms of light...and this manifests itself as the steel gets hot enough to emit light. Red hot steel. Red is light.

So your question is now, if steel is a form of light, why doesn't it glow? It does, actually. We see it; don't we? Anything we can see is only seen because it is reflecting, or generating light. Think about it.

Sight is an amazing thing. If we take optical physics and apply it to everything we see, it is amazing that we see anything at all. Technically, there should not be any sharp, or defining, borders to anything we see. It should all be a blur. The dominant light should drown out the lesser lights of the edges. Moreover, the edges are being reflected in other directions and that should make things blurry. Yet, we see with crystal clarity. Generally.

Our brains have this remarkable ability to take a single ray of light, process it, and reconstruct images which we easily identify. Not only do we see what it is that we are supposed to see, but we see it in three dimensions. If you do not believe that, try lying outside and stare up at the stars in the sky. After awhile, you will have to turn away because the enormity of it is so overwhelm-ing. It is not because there are so many stars...it is because your brain is trying to register the depth of each star. Immense!

Getting back to our piece of steel...it is a block of light in which light is slowed way down. It feels cold. But we wrap an insulated wire around it and apply current to the wire and the metal heats up. It is obvious that we have 'excited' the light of the steel. And, at this point, the steel is now a magnet. It is generating light that is so slow that we cannot see it as radiated light. But we can feel its presence.

If you will recall, earlier I said that there were two kinds of forces in the Universe. Push. Pull (illusionary). Since the Universe is symbiotic, meaning that push and pull are one and the same, the difference then becomes one of direction. For simplicity, let us say that one tumbles through the air from top to bottom and the other tumbles from bottom to top.

The direction in which an atom tumbles is what determines its charge (as we know it). So, if we bring two atoms towards each other, and each is traveling towards the other, then they are going to push one another away. Think of it as two cars meeting on a highway. Ouch!

Now, if you get both cars going the same direction, they simply merge and continue on their way. And so, you ask, if that is so, why is it that the negative pole of the magnet does not attract the negative pole of another magnet?

That one is too easy. You see, as soon as you turn one around to face the other, they are now traveling in opposite directions. Think of it as a flow that is not only going from top to bottom, or vice-versa, but also as going left to right, or vice-versa.

As long as you put the two side by side, they are going in the same direction and are compatible. Spin one around and you have one going left to right and the other going right to left (did you forget the symbiotic relationships?).

I think that the key, the Holy Grail, to Physics lies in the simple concepts advanced by Sir Isaac Newton. Upon reflection, that which seems simple is, in actuality, a pretty complex idea. Let's examine it more closely.

According to Newton, a body in motion will tend to stay in motion until acted upon by an outside force. Since all matter, and all energy, is already in motion, this should be restated as a given quantity (of matter or energy) will act in its customary way until interfered with by another force. And this is true for all things.

Take, for example, a simple toss of a baseball by a pitcher. When it first comes out of his hand, it is accelerating. What this means is that when the baseball is first released from the Pitcher's hand, it is accelerating. At some point along the X coordinate, it reaches maximum velocity. From that point, it begins decelerating.

Effectively, a baseball reaches peak speed at some point a short distance from its origin (release). So the speed of the ball at the very instant that it leaves the Pitcher's hand is not its maximum velocity. You have to understand this concept before you can understand what a Dual Aspect Theory is.

In theory, if there were no outside influences, an accelerating object would continue to accelerate into infinity. In reality, this can only happen if a particle could accelerate to a speed sufficient for it to overcome every influence in the Universe.

As Physicists are well aware, acceleration is the same whether it is to speed up or to slow down. How about an EXAMPLE?

Okay. We know that if we accelerate a human

being, we exert "G Force" on him. Too much acceleration, or G Force, will kill him. If a human being were to travel at 2 miles an hour then, instantaneously, go to 100 miles per hour, that sudden increase in speed would have the exact same result that going, again instantaneously, from 100 miles per hour down to 2 miles per hour. Dead is dead.

I suppose that the hardest thing about this concept is simply to remember that each atom can be influenced by other atoms or forces...but, in the end, it will always have the desire to return to what it was designed to do. It will want to either speed up or slow down. It can be overwhelmed by other force(s) but, since it can never be destroyed, it will eventually revert back to itself.

I am usually fond of pointing to the sky as proof of how atoms behave. I believe that everything that occurs microcosmically also occurs macrocosmically and vice-versa. When I see specific phenomena, such as Halley's Comet, it is indisputable evidence that atoms, or groups of atoms, have a specific assignment. Here we have a comet that comes back every seventy-five years. It has done so for as long as man can remember...or, at least, two thousand years. Does anybody dispute the accuracy of that statement?

IF you see it with your own two eyes, or the majority of people on this planet see it with their own two eyes, can you say it doesn't happen? Of course not. The question for me is not as much does it happen as it is a question of how does it happen?

A Comet has a tail. We know it has a tail because it is meeting some kind of resistance which is pushing against it. If it is encountering a force which is causing part of the comet's mass to leave a trail behind it, why isn't this force slowing it down?

Why does this comet keep returning every seventy-five years with absolute regularity? Why doesn't the resistance slow it down so that it comes back in, say, eighty years? And then 90 years? And, eventually, stop it?

Likewise, each planet is spinning. Why don't they stop? Moons orbit planets and planets orbit suns. Why don't these things just stop? We see all of these wonderful things and never question them.

Of course, there is always the possibility that each of these things are slowing down. If one is to believe Biblical accounts of people living several hundreds of years, then it stands to reason that Earth's days may have been much shorter. Jupiter rotates at a speed which makes its day around 9-10 hours long.

If a scientist were to make the assumption that the Earth was spinning much faster in Biblical days, then he could do some simple calculations which would tell us a great many things. By calculating the rate at which the Earth is slowing down, we could then calculate how much it slows down in 10,000 years. In theory, this would tell us about what speed the Earth was moving when the Earth's magnetic poles reversed.

I would be interested in knowing this information because it should be applicable to an atom. If an atome is sped up by application of heat energy, this would cause a reversal in the atom's magnetic pole and explain a great many things. For instance, when we heat water, it evaporates. This could be the product of atoms reversing poles while other atoms did not. If all of the atoms then became like-charged, they would repel each other. Evaporation.

One thing which I find interesting is to look back at Einstein's earliest thoughts. Albert seemed to be caught up in a struggle to define space and its role in the cosmic scenario. He assigned it spatial as well as temporal qualities which seemed to be best defined in terms of intervals. It is sensible to define thus.

However, after much contemplation, I would be more inclined to eliminate spatial qualities in favor of temporal. If we define space in terms of time, we have the same relationship as matter to energy. Space is a path which can sustain any number of coordinates. Since space is infinite, as far as we know, then it stands to reason that the number of co-ordinates may well be infinite as well.

In this symbiotic universe, matter and energy can

be defined as being opposite ends of the same unit. That is to say, one does not exist without the other. Space and time are, therefore, symbiotic in that one does not exist without the other. Of course, the real absurdity is that energy and matter are the same entity.

We can plot co-ordinates on any given unit of matter just as we can plot co-ordinates on any unit of space. In each case, there is still going to exist a unit of time which represents the distance between each co-ordinate.

It is my belief that space represents the past, matter represents the here and now, and energy or motion represents the future. Time, therefore, is motion and is relegated to the future. And this is true whether it originates in the past, the present, or at some future co-ordinate. Time that has already transpired has already mutated into something else. It has become something tangible and with parameters.

Einstein attempted to place parameters on time by defining it as an interval of space. This, in turn, would enable him to define space as an interval. Space, in and of itself, is not capable of being defined as an interval because it is fluid. It cannot be sliced and diced because that which we would define as an interval flows into the next interval. Space is fluid with no defining or definite parameters.

The other error that Einstein made was to assert

that the speed of light is constant in a vacuo. The speed of light is not constant. It accelerates from the source and decelerates at the destination. Moreover, light is made up of bundles of various wavelengths. Even so-called pure light, such as the ruby red of a laser, is made up of bundles and/or is influenced by other light.

The mass of the photons which make up light, are as varied as light itself. Short wavelength light moves faster than long wavelength light. This is because the mass of the long wavelength light is considerably greater than the shorter wavelength light. Red light and green light move at different speeds. But, because the only thing that can affect light is light (or some humongous body of mass), the different wavelengths of light interact with one another in ways that we are only able to guess at and light travels in bundles.

It seems to me that if we are ever going to derive at anything which even remotely resembles a Unitary Field Theory then we are going to have to substitute time for light. Instead of $E = MC$ squared, we would have $E = M t \text{ (time)}$ squared. This should work fairly well as we consider time to be motion. No motion; no time. The more motion, the more time, the greater the energy in the Mass becomes until, finally, the Mass is converted to Energy.

I could piece all of my theories together and, I am sure, there are people out there who wish that I would just organize them in such a way as to make better sense. But I haven't and I won't (at this juncture). And the simple reason is that I have come to realize that people, some people, want to know what the logic progression was that helped me to arrive at my conclusions. It is my hope that they can gain some insight from this progression.

I am fond of stating that everything that occurs at the subatomic level also occurs at the macrocosmic level. In other words, if you wish to see an atom, just look around. I doubt that an atom is doing anything different than what we see for ourselves. And, in all fairness, if it were, then it would be doing things opposite to what we see (based on the concept of symbiosis).

In closing, let me give you some things to think about. Cosmologists have long speculated that some unknown factor explains why Mercury's orbit is slightly different from the next seven. Mercury has to have a more stable orbit than the others because it is the only planet which does not have its orbital path intersected by another planet.

I predict that scientists are going to discover that the ice-age was not set into motion by a careening asteroid or some meandering comet. What really happened is that there used to be a planet between Mercury and the Sun. As the Sun's mass dwindled, it needed an extra dab of fuel. And so the planet crashed into the Sun. Mercury is next.

If one looks at the planets, objectively, you

quickly notice that the four innermost ones are very much smaller than the four largest. In fact, the larger planets have moons that are bigger than any of the four inner planets.

You might also notice that there are a total of three moons amongst the four inner planets and an abundance of moons beyond the Asteroid Belt. We must conclude that there is more than just a tiny chance that the Earth, Mars, Venus, and Mercury, were all, at one time, moons from either the Asteroid Belt (when it was a planet), or from one of the larger planets. I postulate that they were moons of the planet that broke-up.

When you examine this, it seems very obvious that, whatever calamity befell Phaeton (the Asteroid Belt), the four inner planets were nothing more than moons which were drawn towards the sun at the moment the host planet disintegrated.

It is only supposition, but I assume that our solar system originally only housed five planets. The nearest would have been Phaeton. From Phaeton, each successive planet would have been exactly 2x the distance of the preceding one.

Moreover, we can calculate the speed of the rotation of Phaeton by the following formula: The number of revolutions (rotations) a planet makes in its solar year exactly equals the number of days in the next planet's solar year. I calculate that to be approximately 4254 rotations per its solar year. You can use Keppler's laws to calculate length of rotation per day.

Does an electro/magnetic field invert at set distances based on (mass?) (distance?) (time?)? And is this point of inversion (co)incidental to a given planet's orbital path? Does the shape of an atom assist in determining its properties (as we know them)? What if "Push" really is the only force at work in the known Universe?

PART TWO:

A new beginning…

INTRODUCTION REVISITED

It has been said that any idea which can be proven is, thusly, elevated to a position in science. Conversely, any idea which, largely, remains unproven, gets a minor notation in the world of philosophy. As a direct consequence of this method of classification (segregation), God is placed on the back burner while the lofty atom assumes a primary role in creation.

Not many people, these days, has ever made that comparison. God assumes a position in each of our respective lives based upon a purely arbitrary inclination to accept, or reject, the concept of an Almighty Creator. Atoms, on the other hand, are much touted. And we would be hard-pressed to find even one person (in the so-called civilized world) who would dare to refute the very existence of an atom.

Personally, I find it a strange irony that God has been pushed out of our schools while, simultaneously, the atom enjoys unabated universal acceptance in these self-same institutions. The irony? We can neither see God, nor the atom. Both, logically, belong in the realm of philosophy as we cannot prove the existence of either.

I am not writing this book to prove, or disprove, the existence of God. Nor am I writing this book to prove, or disprove, the existence of the atom. What I *am* striving to do is to get man to do something which he seems to have forgotten how to do. I am, of course, referring to a little used concept called 'thinking.'

So why has mankind, as a whole, forgotten how to think? Well, the short answer is because we have all been brainwashed or, otherwise, duped, into accepting hypotheticals as facts...the existence of atoms being only one of these facts.

I believe that I have the answers to a great many of the questions which have perplexed man ever since the advent of science. However, what I believe, and what I can prove, may or may not be enough to revolutionize the field of science. Be that as it may, I am assured of one thing. Actually, two things.

First, and foremost, I will provoke you into using your gray matter. You will no longer be content just to accept things as your peers tell you to believe.

Second, I shall steer you into the right direction for making new discoveries. I shall do this by revealing what I think is the true nature of the universe. And, as I advance each theory, I will endeavor to reveal its significance, how it might impact some other field, and how we might prove its validity.

As much as I hate to admit it, I have no (as of this time) fully developed and/or useful description of the atom. I am not entirely in agreement with the Bohr model. In fact, all I can say, at this time, is that I feel an inclination to abolish Mendeleev's Periodic Table as being erroneously ineffective and grossly misaligned.

Wait a minute. Hold the phone. Stop the presses. Something's wrong. How can I call my book "Unified Field" theory and, yet, I exclude a depiction of atomic or sub-atomic particles?

Well...whole mule! I never said that I could not, or would not, attempt to construct a model of the atom---assuming that one exists. What I said was that I do not, currently, have a clear idea of how the atom may be constructed. Please allow me to explain this apparent contradiction.

Albert Einstein, among others, felt that there should be a formula, or two, which would explain everything in the known universe. Such formulas should define everything from stellar systems to miniscule Quarks. This, by definition, would be the Unified Field Theory.

By using simple logic, I advance theories which accurately defines gravity, black holes, etc. It therefore follows that such theories ought to define sub-atomic particles. The exception to this hypothesis is just in the possibility that the sub-atomic world may behave entirely different than its cosmic brothers. We shall see.

For many years, I felt I was on the right track. But there was always one thing, in particular, which haunted me. My nemesis was in the realization that, under Newtonian precepts, starlight could not, possibly, travel all the millions of light-years to Earth. Yet, I have witnessed this phenomenon for the majority of my life.

Einstein's nemesis was in trying to define gravity in such a way that it would work for all things, large and small. He could see a cause, and he could feel an effect, but he just couldn't connect the two.

Starlight. Therein lay the riddle of the universe. How could it travel so far, night after night, day after day? I was stumped.

Scientists or, I should say, mathematicians, have busied themselves crunching numbers. You could say, "They cooked the books." Instead of searching for the definitive answer, they provided us with a mathematical fantasy. Starlight, according to them, gets here because stars are really superhot suns.

In a word, bullshit! Starlight gets here because it is really, really, fast---and "True" space is something other than curved space. Yet, even that wasn't the whole solution. A third factor lingered in those dark regions which we often choose to ignore because to accept it would border on the absurd.

Funny. Rod Serling is never around when you need him. Probably off playing some crazy variation of chess with Mr. Hitchcock.

SPACE AS A 3-D ENTITY

Even though I developed my "Push" theory a long time ago, and it solves a great deal of problems, I encountered several more which were not resolved by Push alone. Amongst these problems were the manner in which distant bodies (in space) seemingly acted upon one another, and, also, the way in which magnets seem to be polarized.

The problem of magnetic attraction could be explained by a combination of field and spin. The larger problem of cosmic attraction relied, almost entirely, on field. In either event, the importance of space could not be ignored. Einstein's declaration that the Ether was of no consequence, as far as Relativity was concerned, was a monumental blunder.

Incidentally, it is historical fact that Einstein was opposed to the idea of quantum mechanics. It is also historical fact that Max Planck introduced the concept of quanta five years before Einstein introduced Relativity. How Einstein ever got credit for 'fathering' quantum physics eludes me (not).

Be that as it may, space has become much more than a coefficient, or measure, of time. By my estimation, spatial function(s), or qualities, are much more important than (re)constructing the atom. Indeed, space may well be the nucleus of the atom.

There is a tendency for us to look at space and think of it as being one dimensional (two, if you count time). We have exhibited so much contempt for space that we readily dismiss it as meaningless or insignificant. It is time to go back to the beginning.

Picture, if you will, an immense ball of matter. I mean unfathomly huge. This massive body, according to those who support the Big Bang, contained everything within the universe. Do you know what it did *not* contain?

If you guessed 'space,' you are absolutely correct. Space simply did not exist.

Now, ignite the Big Bang. See how everything flew apart? Big chunks of this. Small chunks of that. And, eventually, very large patches of space.

Are you following this line of thinking? Before the advent of Big Bang, there was no such thing as space…or, rather, large bodies separated by space. After Big Bang, lots and lots of space. Pretty simple, huh? Not really. Let's jazz it up a bit.

Man has a wicked tendency to do one of two things. One, we try to simplify everything by rationalizing it away. Two, we complicate things by irrational thought. Genius, therefore, becomes a matter of connecting the rational with the irrational.

Many theorists postulate that Big Bang was really two explosions which occurred within micro-microseconds of each other. The reality is less complex.

Ever hear of a supernova? Could that not, rightly, be called a 'Big Bang?' Instead of chasing the absurd, shouldn't we be accepting the obvious? Instead of creating a universe from one huge explosion, which could not happen with infinite mass, with infinite energy, and infinite time, why not enlist the idea of many, many, Big Bangs?

Every galaxy began because of a Big Bang. Farfetched?

Not really. I heard, recently, that astronomers are watching while two galaxies collide. Two galaxies! Wow! Can you imagine our sun ramming into another sun? Then repeat that process for every sun in each galaxy. Truly awesome!

Instead of complicating matters, didn't I just simplify them? Actually, no. You see, I just stuck an infinite number of events inside a much larger event which, in all probability, is encased inside a much larger event, ad infinitum.

Each galaxy is a closed system, and must be treated as such. This creates a singularity which has not played out yet.

The Big Bang of our galaxy has created an invisible wall around the galaxy wherein there exists a pressure on all bodies within the system. This pressure comes from all directions at once, from the perspective of the individual bodies, and creates a state of equilibrium (of sorts).

The explosion broke up a large mass into much smaller ones. Space, within the galaxy, is really nothing more than the smaller particles spreading out in ever thinner and thinner layers. In other words, matter and space are manifestations of the self-same entity.

The implications of this theory are as immense as space, itself. Matter and energy become two distinctly separate entities with very different properties. Nonetheless, they are like chameleons which can imitate one another.

An explosion, while composed of finite particles, behaves as infinite waves. This gives the galaxy a distinctive sound as it resonates throughout the universe.

A symbiotic relationship exists wherein energy can only travel where there is matter. This insures that each galaxy can develop independent of the other. In a manner of speaking, this keeps our system from bleeding off into space at too rapid a rate.

Calculating the precise hows and whys of our galaxy should be relatively (sic) easy. But what about the Big Bang? Can we explain that in terms which are more rudimentary to our palatial appetite for comprehension?

In the next chapter, I offer my best guess scenario. There are many possible variations of it. Most of those are inconsequential, and would be better served as classroom experiments.

THE BIG BANG BOOMS

Picture, if you will, a huge monolithic structure floating through space. For ease of understanding, let us say that this huge mass is the Earth (without its atmosphere).

Now we cover the whole planet with an immense blanket of methane and nitrogen. This whole ball is quite cold---close to absolute zero.

To further simplify this, visualize this massive conglomeration with several moons of solid ice. These moons are circling the planet at a nice leisurely clip. Let us give this hypothetical planet a new name. Let us call it Jupiter.

Just to make the experiment more interesting, let's add three similar sized, and similarly constructed, planets to the experiment. We shall call them Saturn, Uranus, and Neptune. Yes, of course, we shall put all four into orbitals around a giant orb we'll call Sol.

Aw, the hell with it! Let's just use our solar system as the example. That way, all we have to do is to introduce some fictitious catalyst. Fair enough?

Somewhere, in the vast darkness of space, there is an asteroid which we shall call the XP5000. The XP represents X-tra Potent. And the 5000 denotes that it is 5000 miles wide. It might also mean that it is traveling at a velocity of five thousand miles per second.

The XP5000 is 50,000 miles long and is traversing through space in a straight (sic) line. XP5000 is an icicle of solid oxygen. It has only one mission in life. It is compelled to plunge deep into Jupiter's heart like one of Cupid's arrows.

As XP5000 pierces through Jupiter's 'atmosphere,' it begins to warm up. At the same time, pressure on Jupiter's surface increases because of the tremendous pressure on it. An explosive fire ignites as the tip of this galactic arrow plunges deep into Jupiter. But that is only the beginning.

XP5000 plunges deeper and deeper. Monstrous explosions are fiercely ripping the planet apart. Then the worst-case scenario happens. The icicle strikes the molten hot core of the planet. The resulting explosion is unimaginable.

Jupiter has exploded into a boggling hail of molten, fiery, globules of all sizes. One of these globules strikes, let's say, Neptune, and it bursts into a fiery inferno much like our sun. In the meanwhile, some of the larger chunks strike the Earth. Some fly into the Sun.

Earthlings are powerless to stop the carnage. All that we can do is to watch. And we do so with the fearful realization that the end is at hand. Disaster looms overhead.

Sure enough, some immense chunks of the stricken planet fall into the Sun. Huge solar flares leap higher and higher; more intense than any we have observed in all of man's history. At this point, whether we know it or not, we have approximately eight minutes in which to say our most fervent prayers.

We are aware, from previous experiences with large solar flare-ups, that these eruptions cause thousands of volts of electric current to flow in the very ground that we stand on. Those events had only taken place for a few seconds. And in fairly localized areas. Our newest flare-ups last for several hours, and cause tens of thousands of volts to flow in virtually every region on Earth. Few people, if any, survive the onslaught.

As the massive electric currents surge through the earth, they alter our magnetic infrastructure. More than likely, this disruption causes the Earth to topple over as its poles reverse. This, in turn, causes the Earth's oceans to wash over all of the Earth. And the Earth wobbles like crazy.

Over the course of thousands of years, amphibians will remember to grow legs so that they can crawl out of the murky water. And apes will, once again, swing from trees.

This was, of course, a scaled down version of a Big Bang. Was it entirely nonsensical? Not at all.

A few years ago, scientists peered skyward and announced that an unprecedented event was going to take place. It was not so much that the event had never happened before, it was just that, on those rare prior occasions, the predicted event was never observed by scientists. And the whole world clambered to watch as a very large meteor slammed into Jupiter. The big event, thankfully, fizzled.

What we were not told was that these educated fools were afraid that Jupiter was going to ignite upon impact. Can you even imagine living in a solar system that, at least temporarily, had two suns?

I had the awe-inspiring privilege of living at a time when Haley's comet soared through the solar system. In addition to being awestruck, I felt the fear that most lesser creatures must have experienced at seeing such an impressive occurrence, or abnormality.

Sadly, the comet blasted through all too quickly and we are left with vivid memories of a giant ball of dirt with a trailing tail. And we were saddened by the realization that it will not return for another seventy-five years.

Very recently, we have learned that two far-away galaxies are in the process of colliding. How long will it be until somebody in that system rises to ask: "Did this all start with a Big Bang?"

A LOOK AT E=MC(squared)

Thanks to a famous man named Albert Einstein, we all have learned to look at the world around us in a very different light. Indeed, a most peculiar and elusive light.

When Einstein introduced his Special Theory of Relativity in 1905, the vast majority of the world simply yawned and then set about ignoring him. Of the remaining few, most were content to lambaste him as a charlatan, a curious little Jew who narrowly escaped confinement in some loony bin, or an outright fraud. To these intellectually deprived denizens of science, Relativity simply could not exist.

Many of Einstein's colleagues refused to delve into the metaphysical realm of Relativity because they did not understand it. And, of the remaining few who did comprehend it, at least on some level, very few dared to wrestle with the complicated equations which the theory represented.

For the better part of ten years, Albert himself would grapple with the complexities inherent in Relativity. Only then could he release a more complete, and updated, version of the theory. He called this version the "General Theory of Relativity."

Through a stringent series of discourses, discussions, and lectures, Einstein's latest version of Relativity was greeted with a tremendous amount of admiration. Even the average citizen (worldwide) got into the act as people squared off on opposing sides of the theory.

As amazing as the theory, itself, was the manner in which people had been duped into embracing it. We all like to think that we are as smart as, if not smarter than, the next guy. When the news was released which professed that less than a dozen people in the world understood the theory, would-be geniuses intimated that they, themselves, fully understood the theory.

Human nature had dictated that men of meager intelligence had to attempt to trick their fellow man by merely suggesting that they understood Relativity. Honest men, many of whom were much smarter than their boastful contemporaries, adhered to the truth. That is, they lacked the training or intuition necessary to perceive such a robust theory.

Man's vanity kept Relativity alive in those early days. Man's vanity continues to keep Physics entombed in a kind of visionary dark ages. In fact, I have, personally, lost count of how many times I have encountered a stone wall of opposition and bigotry. Do you have any idea how many science institutions steadfastly refuse to consider anything which refutes the current definition of Relativity?

Science has not bogged down because of a lack of intelligence. It has bogged down because of narrow-mindedness, tunnel vision, and an unshakable belief in something which, by all accounts, simply does not work. What?

Whenever you talk to researchers, they will tell you that Relativity works well for some things. But it does not work well for others. Where *I* come from, an equation which does not work in all instances, is an incomplete, or erroneous, equation which clearly leaves the door open for a better theory. Yet, the thugs who stand guard over the various science journals, institutions, and what-have-you, refuse to consider anything not in accordance with a hundred year old theory (Relativity)...even though they will admit it is more philosophical (hypothetical) than factual.

How stupid is stupid? Science has, historically speaking, only made advances when somebody has come along and proven everybody else wrong.

In every instance, the collective will of the elitist few has steadily strangled the opposition and refused to open the cognitive door to knowledge. So, whenever I encounter staunch opposition to change, I must question the wisdom of the censors. And I find them greatly lacking.

Remember that old adage? Come on; I know you do. Remember this: Those who ignore history, are doomed to repeat it.

Einstein had his fifteen minutes of fame. It is time to move forward. Question is, where do we move to?

One approach we can take is to dissect Albert's most famous equation. On one side of it, we have an unknown quantity of energy, which we call 'E.' On the other side of the equation, we have another unknown quantity. We call this one 'M' for some measure of mass. The only variable known to us is the speed of light (C). And we square that. Thus, the whole equation is scalar, and amounts to little more than a hypothetical acceleration of 'M.'

That's it. Einstein's most famous equation is for hypothetical acceleration of an arbitrary unit of mass. There is nothing mysterious or profound about that. In every acceleration, be it negative or positive, there is a corresponding change in energy, which energy is equivalent to that which is needed to effect the change in acceleration. More specifically, E=MC (squared) denotes a change in light energy (quanta).

Interestingly, Einstein's equation(s) work very well in the miniscule sub-atomic world, but not so in the macro-cosmic. The explanation lies in the fact that such high speeds are only possible in the one-dimensional, and two-dimensional, worlds. The three-dimensional world is much larger and speeds are, correspondingly, slower.

It is important to keep in mind that these laws are only applicable in a closed system which contains curved space. Our galaxy satisfies both of those requirements. Once we leave the confines of the Milky Way, the rules change.

In true space, the space which lies between galaxies, the speed limits approximate infinity (for light). What this does to the ratio of mass to energy is anybody's guess. Nonetheless, you can be sure that it is extreme.

Just as dimensions have speed limits, at least within the parameters of curved space/closed systems so, too, do moving bodies. Every body which moves, moves in six directions simultaneously. Motion, therefore, is a net force.

For our purposes, dimensions do not exist. We can deal with them in theory but, in our Universe, everything behaves in three-dimensions. On larger scales, it is not practical to deal with one-dimensional matter...although we do tend to view everything in two-dimensions.

The Unified Field, which Einstein so heartily pursued, was within Albert's grasp. Shunning obscure notions of dimensions, and the idea of a light with constant velocity, we are left with a very simple formula. The size of a given unit of mass determines its maximum speed. This is equally true for things as small as photons, or things as large as our galaxy.

Now we come to the exceptions to the rule. Angular momentum. Velocity is a net force, in a specific direction away from a given point (for our purposes). Got it?

Let's look at a sphere. For simplicity, let us call this sphere 'Earth.' Our sphere has an equatorial circumference of just over 24,000 miles. Earth rotates on its axis at just over one thousand miles per hour. Earth also revolves around the Sun at slightly more than 66,000 miles per hour.

While all of that spinning is going on, our galaxy is spinning, too. Since I lack the data, not to mention the computational skills necessary to manipulate that data, I will leave it to the brainiacs to calculate the net force of all of that spinning. Moreover, I will leave it to them to calculate what affect each body has on the other.

Aside from offering scientific proof that Astrology is founded in reality, what is the point in all of this spinning? Scientists, including the illustrious Mr. Einstein, have been acutely aware of the effects of motion within motion. As a matter of fact, Einstein helped to improvise the gyroscopes which the Germans needed to utilize in order for their U-boats to navigate accurately during World War Two.

If we divide our sphere up into 360 degrees, we see that only 1 degree goes in the same direction as the net force. This means that 359 degrees are in directions contrary to the net force. The consequence is that we are more cognizant of the Earth's rotation upon its axis than we are of the speed with which we orbit the Sun.

Gravity becomes a consequence of both the Earth's rotation and its journey around the Sun. Add to that the motion of the Moon, as well as the motion of the Milky Way, and you can derive at a precise formula for calculating gravity.

So what does all of that have to do with E=MC (squared)? Everything in our galaxy, including atoms, is spinning. Everything also has a net force, or velocity. It seems rather strange to me that a formula which excludes these factors could be deemed accurate. Therein, we can easily see why Einstein's famous equation works well for some things, but not for others. As we shall soon see, in the coming chapters, there are even more problems with E=MC (squared).

EINSTEIN GOOFED

Albert Einstein was, without question, one of the greatest Thinkers of the twentieth century. He was, also, one of the most misunderstood.

People tend to think of Albert as a pioneer who came up with revolutionary new ideas which had never been thought of. Truth is, most of the various components of Einstein's theories were already in existence. For instance, "quanta" was a concept introduced by the illustrious Max Planck; a full five years before Einstein.

Even the notion of "relativity" already existed. It had been postulated by men such as Poincare, Lorentz, and others. Again, years before Einstein came along.

Essentially, Einstein merely connected the dots and released his findings in abstract mathematical forms. Hence, his infamous E=MC squared. But Albert's thinking, like that of his predecessors, was flawed.

Early in his career, Einstein attempted to prove the existence of something scientists called the "ether." This ether was supposed to be some unseen substance in outer space which would allow, or enable, light to propagate from one end of the Universe to the other. Without it, they speculated, light would not be able to move from one place to another.

By the time Albert formulated his theory of Relativity, he had concluded that the existence of the ether was so insignificant and inconsequential as to render it unto the realm of "who cares?" Oh-oh!

In my humble opinion, "proof of the existence of the ether was right under his nose. In fact, it was the ether which was influencing Albert's theorems."

I suspect that the ether runs at, or very near, the speed of light...but, in the opposite direction. "Think of it this way: A light quanta travels to its destination, drops off its cargo, and then returns towards the source to replenish."

For more than a century, scientific researchers speculated that light had no mass. The photon, in particular, cannot have mass because, if it did, it would smash us to smithereens. I assert that a state of equilibrium exists which nullifies the potentially harmful affects of light quanta.

Einstein spent the greatest portion of his life searching for the Holy Grail of Physics: A Unified Field Theory. According to my calculations, "It is ironic that Einstein abandoned the one thing that would have enabled him to piece the puzzle together. The ether is real!"

Because the ether is a component of a system in equilibrium, it is virtually undetectable. If we push X amount of Ether ahead of a moving object, X amount of Ether will move behind it in the same direction of travel, and at the same magnitude.

"The Ether exists and I can prove it." I can say without hesitation. "And I will explain how you can prove it, too."

The method of proving it is the topic of another chapter. So I will leave it at that for now.

THE ARGUMENT FOR ATTRACTION

It is pretty easy to argue the case for attraction. We look at phenomena such as the forces exhibited by magnets, and we feel that we have no choice but to conclude that the force of attraction exists. And, yet, I am still not convinced. Not even the apparent effect of the Moon upon the tides can sway me.

However, I am perfectly willing to play the Devil's Advocate and argue in favor of attraction. Let's see how that pans out.

The forces of attraction and repulsion imply a relationship such as negative-positive, push-pull, or any similar force relationships. Attraction, Negative, and Pull, all suggest a void wherein atoms can migrate. That is the simplest scenario.

In this 'void' scenario, we have microscopically big canyons in which certain atoms can fall. Still, in order for such a scenario to exist, both the entrance to the 'canyon,' and the atoms spilling into it, must be smaller than other atoms. It is plausible. And it lends credence to the possibility that negative numbers could have physical manifestations within our Universe.

Obviously, if we cannot advance a satisfactory argument favoring the force of attraction, even with the aforesaid simple scenario, then something else must be going on.

Numerous scientists have attempted to explain the force of attraction, at least in some gravitational sense, throughout the ages. To the ancients, magnetic attraction, via lodestones, was magic. To later investigators, such attraction was the result of properly aligning the bipoles within a given magnetic unit. Modern researchers toy with the idea of atoms throwing dust at one another.

My personal favorite is String Theory. Envision, if you will, my 'advanced' version of String Theory. Let's call it "Z Best Theory."

In Z Best Theory, we take microscopic pieces of string and hand each end to a person standing on an atom. Thus, two people float around playing a fastidious game of tug-of-war. All day. Everyday.

Just as their real-life human counterparts, these little guys get tired and they drop the string. A new (rested) opponent then picks up the trailing string and a new battle ensues.

This silly discourse is, logically, repulsive (sic). Nonetheless, it does open the doors for serious speculation. If we eliminate the tiny string holders, and substitute some, as yet, unknown holders/fasteners, the idea of tethered atoms is not nearly so preposterous.

Carrying the concept of tethered atoms a step further, we can assign differing qualities to each of the tethers. One can be stiff like wire. Another could be flaccid like rope. Still another could be flexible like a bungee cord. Perhaps we can call these thrill seeking atoms "bungee jumpers!"

To this whole conglomerate of tethered atoms, we can add in other factors. Again, these qualities would only apply to the tethers. For instance, we can vary the lengths so that two larger atoms could only 'duke' it out if they encountered a longer tether.

On the surface, it would seem that this fanciful version of string theory would be readily acceptable to even the most staunch supporter of either string theory, or of the force of attraction. The inherent problem is, how do we define the contact points for the strings? Are they held in place via ancient magic? Or is there a force of attraction at work in ways that are unidentified by us?

I must say, with some relief, that I have failed to advance a proper argument favoring attraction. It is possible that I failed because I was wanting to fail. On the other hand, maybe I failed because attraction is a myth. Whatever the reason, I'm sure that you won't want to miss the next chapter.

EINSTEIN'S SECOND MISTAKE

We are all so conditioned to believe that everything in the known Universe is symmetrical that we obscure our vision and our intellectual capabilities. This symmetry is a logical conclusion based on what we can observe. We see north-south, up-down, negative-positive, etc., etc. But symmetry is an illusion.

Each pairing is merely the opposite ends of the same thing. Instead of being two disparate functions, these are all the same. But they appear opposite because we view them from opposite ends of the event(s).

In the most revolutionary discovery of the twenty-first century, I have concluded that, "Push" is the only force at work in our Universe. "Pull" is only an illusion which simply does not exist in the three-dimensional world. It is an elementary concept which eluded Albert Einstein in that great man's quest for a Unified Field Theory.

Back in Einstein's day, scientists were too busy squabbling over the existence of the atom to take a closer look at forces. Only a handful of men were brave enough to actively promote the atom. I predict that my theory will meet with the same initial resistance. Still, I am confident that my conclusion will be vindicated.

To illustrate the degree of opposition I expect to encounter, I point to another belief that has long been held onto by Physicists. That belief is that nobody "outside of the profession" will ever receive a Nobel prize for his work. Wanna bet?

If I am correct, then I have given researchers the missing ingredient for the Unified Field Theory. If that theory is the chief goal of Theorists, how will they ever justify excluding an "outsider?"

Right, or wrong, one thing is for certain...life just got a whole lot more interesting.

REDEFINING GRAVITY

Early on, I had come to believe that I was on the verge of proving that Einstein goofed as he assembled his theory of Relativity. I had cogitated the idea of eliminating the force of attraction several years ago. And I knew that it contradicted much of Einstein's work. Just how true that was did not surface until I made an even more startling discovery.

Having formulated my "Push" theory, I went in search of proof which would substantiate, or refute, the idea. I looked at the moon and realized that it could not be attracted to the Earth. This, however, was not an entirely knew idea to me.

I had always been fascinated by magnets. I knew that "like" poles, or charges, repelled. Opposites attracted. It was these simple rules which I always applied to the moon.

The moon reflects sunlight. That could only happen if the moon absorbed enough energy from the sunlight to become saturated. The Earth, also, reflects sunlight. Again, this could only occur if the Earth had become saturated, too. Obviously, if we have two bodies which contain like charges, then they are repulsive to one another. Extremely logical.

So far, that only utilizes classical Physics. Anyone with a minimal of intelligence can see that.

Like millions of other people, I embraced Einstein's Relativity and applied it to the Earth-Moon relationship. I had no problem with putting both of them inside a curved space that wrapped around them like taut cellophane.

From there, I toyed with the idea that space was compacted in the areas all around each orb. It is a hypothesis which I only slightly veer away from today.

In my mind, there are two types of space (perhaps more). There is the space that needs to exist just as an artist's canvas. Without this stationary space, there is no 'art!'

The other space, the one I call the Ether, is much like a succession of gears. It moves with matter.

This space would best be described as semi-rigid. And it moves at right angles to primary space. It is this space which curves and gets compacted (denser) in places of greatest activity.

I would almost be tempted to call these two spaces electro and magnetic. In that way, I could assign them very real attributes. Moreover, such definitions would alienate them from the dust particle space which permeates all regions of a closed system.

It was very easy for me to see that the moon was not attracted to the Earth. And vice-versa. Still, I wondered about the existence of any other proofs.

Introduce Pluto and Charon. One is (or was until recently) called a planet; the other, a moon. The fact is that many Astronomers do not, and have never, considered Pluto to be a planet. I add myself to that list...kind of.

Pluto and Charon are an interesting phenomenon. They are, essentially, the same size---more or less---and go through space tumbling around one another. It is a beautifully strange dance at the outer fringes of our solar system.

I had found my second clue. Pluto and Charon could join together...if they were attracted to each other. But such a marriage would never see a honeymoon. They can never join as one.

Without a doubt, if a force of attraction existed between the two, the curvature of space around them would add to the attraction. I had found my second example in support of push.

One thing that I had, inadvertently, overlooked in my efforts to resolve Einstein's mistakes, was the simple fact that I was really trying to define gravity. And that was the precise instant that the gravity (sic) of the situation hit me. I wasn't just trying to define gravity; I was rewriting the fundamental equation with which so many learned men had devoted their entire lives.

Isaac Newton had defined gravity as a force of attraction. Every great man who followed, blindly accepted Newton's definition as gospel. Whereas I had merely intended to expose the error in Einstein's attempts to link gravity to relativity, I had really, in actuality, gone all the way back to Newton.

And Newton was wrong. Boy-oh-boy. His formula for calculating the force of gravity had worked so well (for many purposes) that nobody ever thought there could be another answer. Until now.

Specifically, Newton said that the force of attraction between two bodies (usually planets or moons) is directly related to the mass of those bodies and inversely proportional to the distance between them. What that meant was that bigger bodies would have more attraction than smaller bodies. Just what that force was, nobody could say for sure.

One of the most obvious problems inherent with Newton's hypothesis was that it did not make any difference what these bodies were composed of. All that mattered was their physical size.

Think about that one for a moment. A giant ball of steel would attract an equally massive ball of glass. In fact, according to Isaac, the two objects could be any shape and, therefore, did not need to be round, oval, conical, or spherical. They could be any shape!

Well, you really don't need for me to tell you how crazy such a postulate is. So I won't. However, I do feel a certain compulsion to explain the most plausible reason why Newton's theory was so widely accepted---particularly by advocates of the Relativity theory.

Einstein speculated that gravity was a consequence of space warping. In other words, the larger the body, the greater the space around it warped. That was a very noble attempt at explaining the phenomena. And there may be a thin element of truth to it.

Now you must be wondering just how I propose to explain gravity in such a way as to eliminate the force of attraction, and still allow for the components of Newton's theory. Actually, that is quite elementary.

Our galaxy is the result of a massive explosion. No, not the Big Bang. Much, much, smaller. But large, nonetheless. And the fact that the Milky

Way is spiraling we can deduce that the explosion was the result of a major collision between a couple of suns, or so. This means the suns collided slightly off-center to one another.

Be that as it may, the force of the event is still contained within our system. My best guess is that explosions in space occur in slow motion...and last for a rather long interval of time. That is, relative to us, it is a long event.

The force of the explosion, possibly as some form of heat wave (cold wave?), bounces off of the outer walls of the galaxy...and everything within the galaxy. Because it is a closed system, there is an equal magnitude of force being distributed throughout the galaxy.

Though this force is spread equally, it does not act that way. Let me give you an example. Let's say I put a golf ball in the very center of a pressure cooker and seal it in. As I pump air into the cooker, the golf ball should remain stationary.

Now, let us open the pressure cooker and place a second golf ball inside. We seal the cooker and begin to add pressure. Eventually, the two balls will collide.

Next, we put a baseball in the center of the cooker, place a golf ball somewhere in the vicinity of the baseball, then add pressure. Only this time, we sit the apparatus on a Lazy Susan and spin it around. If we do it right, we should observe the golf ball 'orbiting' the much larger baseball.

Given enough time, it would not matter where we placed the two balls originally. Over time, the much larger ball would always find its center of gravity.

Two bodies in space are under the same sort of force. If we draw a line, a straight line which begins at the outermost edge of the galaxy, passes through both bodies, and ends up at the outermost edge of the other side of the galaxy, we will make a simple discovery. The space between the two bodies is shorter than the space between the bodies and their respective galactic walls.

LOOKING AT DIMENSIONS

One of the greatest blunders of the twentieth-century was placing *time* in the fourth dimension. We have all been taught to place things in their order of importance. This might explain why the fourth dimension has been so badly downtrodden or misunderstood.

According to Atkins, "Time is everywhere. If time did not exist, nothing would exist." Touché!

Because time permeates everything in the known Universe, it should be relegated to its rightful place in the first dimension. Next would come space because it is the stage from which all actors act. Third would come light. Fourth would be electricity. Fifth would be magnetism. And sixth would be matter. Why?

Time is infinite. Space can be infinite or, as many suggest, "finite, but without borders." From the infinite, we work our way on down the speed limits.

The first observable phenomenon, which has an observable speed, is light. Light is a one-dimensional entity which sets the speed limit of the first dimension at about 186,340 miles per second (mps) in our galaxy. Atkins intends to prove, in another paper, that the speed limit for light is near infinite outside of our galaxy.

Going on, electricity is at the border between one-dimensional and two-dimensional. Electricity travels at about 186,000mps. At the other end, lies magnetism. Magnetism is at the cusp between the second dimension and the third. It travels at a much slower speed than electricity. Both electricity and magnetism are two dimensional entities.

Matter, or three-dimensional objects, are much slower. These speeds are regulated by states such as Gas, liquid, or solid. It is important to remember that, in no known dimension is there anything that is purely at rest (relative to the known Universe).

From the aforesaid facts, according to Atkins, we get a dramatically improved picture of the relationship of energy to matter. Effectually, energy is matter in motion, and matter is energy at rest (but not zero).

BLACK HOLE OBSERVATIONS

In mathematics, we often utilize nonsensical expressions which are then applied to explain hypothetical situations, or events. Negative numbers, and fractions, immediately spring to mind. And the saddest part is that all of us have developed thought processes which enable us to grasp these abstract concepts.

Let's say that I have a stack of ten silver dollars. To this, I wish to either add a negative five silver dollars, or I wish to multiply by the fraction 1/2. What do I wind up with?

We, of course, conclude that the outcome is that I now possess half as many silver dollars as I started with. The reality is, there is no such thing as a negative five silver dollars. Likewise, if I multiply 10 silver dollars by 1/2, I do not end up with five silver dollars. What I do end up with is 10 silver dollars that are now 1/2 the size they were.

Let's look at another example. Let's say that I have a silver dollar. To this, I wish to add a negative twenty silver dollars. In colloquial math, we deduce that we have a net result of -19 silver dollars. It is an abstract result which cogitates such distressful notions as debt, deficit, and being in the red.

Okay. We've got the basic idea. So what? Well, now we substitute. Instead of silver dollars, let's say that I have a Moon. To this, I add a negative 20 moons. Now what?

If, indeed, there were a force equivalent to a negative 20 moons, we might expect to see a black hole where our Moon used to be. Strangely enough, we should expect to get our Moon back by adding 20 moons to the black hole. Possible?

Who knows? Such hypothetical preponderances may occur in reality. Or they may just be fanciful musings. If they exist, then it is possible that parallel Universes exist. But not in the conventional manner of thinking.

So much mystery and hoopla has been advanced, as regards the black hole, that it almost seems criminal to demystify them. Irregardless, science must move on.

Every galaxy in the Universe is spinning. This signifies that the space within those spinning clusters is curved. Light, traveling throughout these galaxies, is moving at a rate of speed relative to the rate of the rotating galaxy or, if you wish, at a rate proportional to space density/curvature.

I will come back to this spinning, as well as its effect on light, in the next chapter. Let's suffice it to say that all of this spinning slows down light.

The point to be made here, is that there exists two distinct types of space. We have a closed system in which space is spinning (our galaxy). And we have an open-ended space which is at rest (relative to our galaxy). All scientists know that things behave differently when they are at rest than when they are in motion.

In the case of the spiraling galaxy, light travels slowly through the curved space. When light exits curved space (our galaxy), and enters the open-ended space beyond the Milky Way, it rapidly accelerates. A black hole, as well as all black matter, is nothing more than places where curved space has given way to true space.

THE SPEED OF LIGHT IN CURVED SPACE

Scientific researchers have been dogged for more than a century by the concept of relative space wherein the speed of light is constant. Likewise, these Theoreticians have been mystified by phenomenon such as the aberration of starlight. We can, at last, alleviate the apprehension brought on by those perturbations.

For many years, John Hildreth Atkins stared at the night sky in sheer disbelief in what he was seeing. He wondered, as had many before him, how it was even possible for starlight to reach the Earth. That light, he theorized, came from such a vast distance that it should have dissipated or, at the very least, been disrupted or interfered with long before completing its journey.

Then one night, John watched intently as a manmade satellite approached a not-particularly-bright star. For some inexplicable reason, he speculated about what would happen when the two objects converged. While ruminating, the event passed, and that created considerable turmoil over what had been seen. Instead of blocking out the light from the distant star, the craft seemed to veer out around it in a rather wide arc.

The light from that star was clearly "bundled." Yet, it defied logic. How could light, which originated so far away, retain a magnitude sufficient to "push" aside a solid object?

For Atkins, there was only one viable conclusion which would satisfy each of his questions. The solution was that light from those stars was traveling at speeds far greater than those observed in classical physics. But how could that be?

Atkins speculated that light was one-dimensional. As long as it could travel in a straight line, its speed would approach infinity. However, in a closed system, such as the Milky Way, it was forced to traverse along a curved path through space. Aha!

Even if we ignored the simple math of angular momentum, or that of inertia through mixed mediums, we could not ignore the flirtations with other dimensions. Light travels slower in curved space because it is being forced into an additional dimension. And a given unit travels considerably slower in two (or more) dimensions than it does in one.

Some of the most profound discoveries are those which were chance encounters. Put another way, if you went out in search of gold (ore), and you did not know anything about gold, how would you know whether, or not, you had succeeded?

The answer, of course, is persistence. When everything else fails, try again. Always, try again.

THE ABSORPTION THEORY

The entire Universe works on a principle called Cause and Effect. It has, at its roots, another principle which states that Necessity is the mother of invention. As a matter of recourse, we could say that necessity lies at the heart of any precluded Unified Field Theory. And the reason for this is simple enough. Everything exists for a reason.

For those of you who ascribe to Biblical intonations, we can turn to Ecclesiastes for confirmation. Everything has a time and a season. And the inference we can draw from this awareness is this: A thing exists for that time and for that reason.

Another inference we gain from Ecclesiastes is that all things are important. Take, for instance, a strawberry plant. For eleven months of the year---nothing. Then in the twelfth month, gloriously, the plant yields succulent red berries which we can preserve for consumption in any of the months to follow.

Point is, for all of our collective intelligence, we can see many things, yet we do not see their purpose. Let him with understanding, understand; him with wisdom, to be wise; and those without a clue...just shut the hell up!

That wasn't very nice. But oh so true. I am so tired of so-called professionals telling us to believe in this or that, and then they admit that they don't know.

Yes, I am taunting the face of hypocrisy. I ramble on, reaching, it would seem, for straws. And I confess that I am ignorant as to all things. In that context, I am as guilty of misdirection as my fellow man.

One difference between what I seek to accomplish, and what the others are doing, is that I am trying to reprogram your computer (brain). I challenge you to buck the trend. All great accomplishments have come from those who swam against the tide of conventional thinking. Dare to be different. Question everything.

One of the things which I have gained a reputation for doing is to lead you down a path which seems to have little, if anything, to do with the topic at hand. The short answer to that is that the complex nature of my mind sends me down all of these paths simultaneous to arriving at a conclusion. The odd thing is, these excursions into unrelated areas often turn out to be highly applicable after all.

The topic before us is 'Absorption Theory.' What can the mistaken beliefs of researchers, and your adherence to those mistaken beliefs, have to do with absorption? For the answer, let's define the Absorption Theory.

In the exciting world of electronics, there is a wonderful process called saturation. Essentially, saturation means that a unit of material has "filled up" with energy and can no longer absorb additional energy. In electronics, this energy is usually a flow of electrons.

Technicians have a pretty good idea how many electrons a given unit can "absorb" before reaching the saturation point. This knowledge enables the technician to construct various circuits which function a certain way when the given unit is in a particular state, or the other. That is to say, as long as the given unit is 'thirsty,' one circuit will remain in operation while the other(s) are not. As soon as the unit is full, a second and/or third circuit takes over and the electrons are diverted away from the unit.

Astrophysicists tell us that every sun in the observable universe exhibits absorption lines which they call 'Fraunhofer Lines.' Basically, these are dark lines which show up in spectral analysis of sunlight. They appear as dark lines because there is no light in them. Think of them as shadows.

According to Physicists and Astronomers, these dark lines are caused by cool gas (high above the sun's surface). They claim that this cool gas absorbs only certain wavelengths of light and, thus, enabling researchers to identify the composition of these gas clouds by the absent wavelengths.

The prevailing theory about why this is so, is that certain elements absorb specific wavelengths. The other side of that equation is that certain elements emit specific wavelengths that other elements do not.

It seems to me that if the whole spectral analysis thing is so accurate, why don't we eliminate the Mendeleev Table and replace it with a color coordinated chart? We'll divide the chart in half. On one side, we place light emitters. On the other side, light absorbers. And, just to top it all off, we'll demonstrate how the light absorbers become light emitters---once they reach saturation point.

Such an absorption theory (table) is not outside the realm of possibility. And it does make a fair modicum of sense. Hmmmmmm…

PROVING THE EXISTENCE OF ETHER

A hundred years ago, it was hypothesized that, if an Ether existed, we should be able to measure it in some way or fashion. Indeed, the notion that a stationary Ether existed fit very well into various theories and equations. Unfortunately, all attempts to prove that an Ether occupies space have failed---until now!

Convinced that an Ether exists, and equally convinced that light should travel much faster than a paltry 186,340 miles per second, John Hildreth Atkins has formulated a plan that he believes will prove that the Ether is really there. And, almost as a side-bar, John's plan will provide us with a clue as to some, if not all, of its characteristics.

So what, exactly, is Mr. Atkins' proposal? Placing a laser on the Space Lab, and firing it at mirrors on the Moon (or beyond). By pulsing the laser at diminishing intervals (I.E.- one second spacing between shots, down to 1/100 second pauses), we should see an erstwhile response.

Explanation: If light is faster than 186,340mps, and this slower speed is a consequence induced by the Ether, then there is a good chance that the intervals between shots will disappear once the "faster" light catches up to the pulse in front of it. This will occur when the shots occur at a rate faster than the Ether's ability to "mend."

FUTURE TIME

Just a quick note...I was going to write an article/chapter about future time. This being the speed of time and its affect on the way we interpret the universe. It stems from my belief that we are in the midst of "fallout" from a huge explosion and that caused time to slow (as compared to our ability to comprehend). At any rate, I will deal with it at a future time. Smiley face.

ESCAPE VELOCITY

Scientists have done a wonderful job of calculating escape velocities for the Earth and Moon. But why haven't they done the same thing for all things (including the atom)?

My first impulse is to answer my question with a vague generality. I could simply say that such calculations are much too complex. However, in this day and age of computer literacy, it is merely a matter of programming.

If we possess the empirical data, and provided we have the correct formula, we should be able to predict for all objects. But, alas, we cannot. And the reason is simplicity, itself.

Let's take a sphere and half fill it with sand. Next, we fasten a tether to it so that we can sling it around in a circle. We now note that as we increase the speed of our orbiting orb, as it circles us, the sand begins to gather at the outer boundary away from the fulcrum (us).

Do you see the picture? Our orb has an escape velocity. If we spin it around us fast enough, it will fly off into space. But, is speed the only factor? Of course not!

The strength of the tether, the strength of the fastener, and our grip on the other end of the tether, all come into play. Earlier, I referenced that motion was the net force of the total forces at work. I also stated that a moving object moves in six directions simultaneously. I could carry that postulate further and simply state that all net forces are the result of, a minimum of, six forces. Escape Velocity, thus, depends on the mass of the globe (including the sand), the speed at which we accelerate it, the strength of the tether, the strength of the fastener, the strength of our hand, and the forces which make up the ambient.

After our orb has reached the saturation point (I.E.- the point where it cannot take on any additional energy), any additional energy (you can substitute mass) will cause it to 'escape.' With this escape there is an accompanying moment of hesitation, which is followed by a resulting inertial acceleration. In other words, once the threshold of Escape Velocity has been compromised, one of four things can occur.

According to Newton's laws, our orb will continue to accelerate until it is acted upon by an outside force. Therefore, our four choices become: It can stop, slow down, maintain a constant speed, or it can speed up. Again, these actions are contingent upon ambient, or lack of ambient, forces.

The Big Picture

As much as he was able to at the time, Albert Einstein attempted to build a mental imagery of the Universe around him. Such a feat was tantamount to looking at a microbe with a microscope---not impossible, but darn near.

What dear Al ended up with was a model wherein the Universe is unlimited, but is finite in volume and in all directions. We can liken this concept to a balloon which provides us with a visual two dimensional surface within a three-dimensional system. If we were on the surface of that balloon, we could walk forever, in any direction, and never encounter a boundary. The balloon, like the Universe inside of Albert's mind, contains a set volume within a clearly defined area.

When Einstein pieced together this particular concept of the Universe, he was under the impression that the whole Universe was curved and, therefore, the light from a distant star, light that had sufficient energy with which to make such a lengthy journey, would be visible in two positions within the parameters of his Universe. The light would be visible in the relative location of the origin of the light and, also, in the opposite position.

This dual aspect theory could best be illustrated by the following analogy. Let us draw a point on the surface of our balloon and label it a star. Now, we can draw a line around the circumference of the balloon. This we can label the path of a single ray of the starlight. Next, we can point to any position along that line and we observe that, no matter where we select our imaginary point of observation, there is a line in front of us and a line in back of us. This, allegedly, confirms that our starlight is visible in two opposite directions.

The only drawback to the above-referenced experiment is that such hypothetical thinking does not adequately represent infinite space. However, the balloon is an excellent purveyor of a galactian model. Think of it in terms of the balloon being representative of our galaxy, the Milky Way, and start to spin it. Have a couple of million friends or associates come over and hand each of them a balloon. Tell them to start spinning their balloons. Now, we have a very, very, good model of the real Universe around us.

Fact is, we live on a planet that is spinning inside of a galaxy which is also spinning. We, therefore, must deal only with moving forces. Everything inside of our balloon is moving in an arc. Everything outside of our balloon moves in a straight line.

If you have ever looked at the night sky and wondered how the light from a star, in a galaxy which is a million light-years away from us, actually reaches us, you now have your answer. Light, which travels in a straight line between galaxies, travels at, or near, infinity. But how?

In our galaxy, space is curved. We can use Newtonian physics to calculate the effects of inertia, angular momentum, etc. Those with extreme intellect, can see that curved space is compacted (dense). Light travels through dense mediums at a much slower velocity. Moreover, in open (non-curved) space, no dimensions exist.

Let me put that into a different perspective. Quantum physics tells us that a three-dimensional body, with a given mass "M," traveling at one hundred miles an hour, will do more damage than a two-dimensional body with the same mass and speed. This is because when the two-dimensional body (essentially a photograph) encounters the three-dimensional body (all things equivalent), the two-dimensional body's mass will be redistributed in the three-dimensional. The result is a lessening of 2-d's force.

Let's look at a classical equivalent of that example. Let's say that we have three batteries. One is a four volt, one is an eight volt, and one is a twelve volt.

Now, let's suppose that we have three dimensions. Each dimension is capable of holding a maximum of four volts. Let's start out with the four volt battery and place it in the first dimension. It zips along until it encounters either the eight-volt, or the twelve-volt, battery sitting comfortably in its own little sector of the universe.

At the time that the two batteries collide, several things can occur. The two batteries can zip off in new trajectories, they can fuse together, they could altar one another, or they can unite as one (3-dimensional) battery.

One thing to bear in mind about the last postulate is that the 2 combined batteries are not equal to the singular twelve-volt battery. The united batteries can either be greater than, or less than, but not equal to the singular battery. What?

To better visualize this, let's take an excursion into carpentry. In carpentry, it is often necessary to "laminate" boards to one another. In some instances, this creates a compound that is, by itself, weaker than a single board of similar mass. In most cases, the process of lamination is to reinforce, or to make stronger.

The point of this whole exercise is to show that a three-dimensional object has more energy than a two-dimensional, the two-dimensional has more than a one-dimensional, and the one-dimensional has more than space (0-dimensional). For the record, curved space is two-dimensional, light is one-dimensional, and plain old space is 0-dimensional (for our purposes).

Therefore, our one-dimensional light slows down in 2-d space. It can only travel at 186,282mps...until it enters non-curved space. Since 1-d is a force greater than 0-d, light can now travel unimpeded. And continues accelerating at C squared (exponentially).

Our next problem appears when we attempt to travel. Three-dimensional matter cannot travel as fast as 2-d, 1-d, or 0-d, matter. But, even before we can address those problems, we have other hurdles to clear.

Virtually every human being alive today is aware of a little phenomenon called Escape Velocity. There is an escape velocity for the Earth. There is an escape velocity for the Galaxy. Matter of fact, there is an escape velocity for every single atom in the Universe. But let's not hasten there right this minute.

The question now arises, "Why do we have escape velocities?"

Aside from the fact that God designed it that way, we have stumbled upon a fourth dimension which we shall call "Time."

The problem that we face when dealing with the concept of time is that we do not have a uniform definition of the concept. Indeed, for most of us, "freezing" time implies halting forward progress which, in turn, implies that time is motion. Oh-oh!

Einstein's famous equation banks on two postulates. First, the notion that the speed of light is constant. And second, time itself, as related to light, is constant. While the speed of light, within the framework of our "closed system" galaxy, is close to constant, time, most definitely, is not. Let's look at some practical examples.

Let's say that we get into a car and we travel at the rate of one mile per hour (mph) down a given highway. It is a hot day and we have the wife and kids in the car with us. I can pretty well guarantee you that, at the end of one hour, your family is going to make your life more miserable than it already is.

Now, let's return to that same stretch of road. Instead of one mile per hour, let's go 60mph. At the end of one minute, we have already traveled the same distance as our earlier venture. Better yet, is the fact that nobody in the car has gotten cranky or miserable.

The argument arises that we have merely accelerated time. That is horribly erroneous and grievously flawed. For time to have been accelerated, we would need to replicate every event within the 1mph drive. Did we overheat? No. Did we get bored? No. Do you see the point?

Time is not strictly about motion. In fact, if we froze the Universe, time would continue because the universe would still be there. Why? Because potential energy has a time coordinate (coefficient).

Some theorists will argue that if the universe were to freeze, it would cease to function (exist) and time, would also, be nonexistent. Nice theory. But rejected and rebuked by both the Conservation of Matter, and the Conservation of Energy.

An interesting hypothesis arises out of this abstract thought. If we did have a situation wherein the Universe froze, it becomes obvious that we would need the addition of a single moving particle in order to put it into motion again.

I use this abstract to make numerous observations. First, the particle would be the reverse equivalent of a Universe Escape Velocity. The calculation for such a particle would be of a magnitude comparable to the alleged Big Bang theory.

Ignoring the speculation about where such a particle may have come from, as well as what may have propelled it, let us look at a more likely scenario. Let's consider Brownian Movement.

As we are watching the jitters of suspended molecules, which are reminiscent of people who have imbibed too much caffeine, we realize the huge energy potential which is just waiting for liberation (a direction). The liberating particle, say a beam of light, enters and these Brownian Ballerinas spin away.

Physicists know that a good explosion depends on the energy potential, as well as the time (duration) of the release of energy. The nearer the magnitude the potential energy can come to the saturation point of time, the greater the magnitude of the explosion.

We can go back to our car experiment and substitute a bulldozer for the car. We drive the bulldozer at the rate of one mile per hour, twenty-four hours per day, for one full week. We push pile after pile of dirt into a humongous mountain. At the end of the week, we have one hell of a hill in front of us.

Now, let's change the time coordinate. We create the same mountain, but we construct it in one ten-thousandth of a second. What do you suppose happens?

Well, assuming it were possible to even do that, we would probably have a crater in the Earth about the size of the Grand Canyon. The reason for that is simple enough. Heat from things like friction, exhaust, and pressure, would not be able to dissipate. The heat would accumulate very rapidly and the net result would be a huge explosion.

PART THREE:

The Future…

DUAL MASS THEORY

I have always said that the way to discover how atoms work is to take a good look at the solar system. So I listened to myself. I took a long, hard, look at it. And I was shocked at what I discovered.

Planets were not content just to have moons (or not). They had to have companion planets. That's right. Excluding Mercury and Venus, every planet in the solar system has moons. And these planets also have a twin (in a manner of speaking).

Ever take a good look at the statistics? Earth rotates on its axis once every 24 hours and 4 minutes. Mars rotates on its axis every 24 hours and 37 minutes. Pretty close!

Jupiter rotates once every 9 hours and 55 minutes. Saturn does it in 10 hours and 39 minutes. Uranus is 17 hours and 14 minutes. While Neptune is 16 hours and 7 minutes.

Most scientists and researchers, myself included, tend not to pay a whole lot of attention to Pluto. It, after all, looks more like a pair of moons. But who am I to argue. If it is a planet, fine. That means that it is the best example of this planetary pairing in that it flip flops with Charon, its so-called moon. So that makes it the most clear-cut example of planetary pairing.

Before launching into a tirade about why the others are all paired, let me offer an explanation as to why Venus and Mercury seem to be odd balls.

Venus is only fractionally smaller in diameter than the Earth. This is noteworthy because all of the other pairs are very close to the same diameter. More on that in a minute.

Mercury is only slightly bigger than Earth's moon. This leads me to believe that Mercury is Venus' moon. I suspect that if Mercury were orbiting Venus, then Venus, and not Mars, would be Earth's twin. And Venus would be rotating on its axis at pretty close to 24 hours instead of 243 days (in the wrong direction).

Is there any data which would support this theory? Oodles. Take the following calculations, for instance. The diameter of the Earth is about 12,756 km. The diameter of the Earth's moon is approximately 3,476 km. That adds up to 16,232 km. Right now, Mars and its two moons are about half the size of the Earth (alone). So we know that

Mars is not the twin for Earth. So let us look at Venus.

Venus has a diameter of about 12,103 km. Mercury has a diameter of approximately 4,878 km. The total for those two objects is 16,981 km. Take away the Earth/moon total and we have a difference of only 749 km. Pretty darn close!

For the record, the margin of error can be resolved by assuming that the precise measurements have been miscalculated. Alternately, the orbits of the moons can make up the difference in total mass/energy.

A third possibility may be that, in pairing, one of the planetary pairs has to be slightly smaller so as to enable one to lead and the other to follow. More precise calculations of the entire solar system (pairs) would probably reveal this…if the discrepancy is by the same amount in every pairing.

How would this Earth-moon/Venus-Mercury pairing compare to the other planetary pairs? Let us look…keeping in mind that Mars is an imposter. Mars, like Mercury, is a moon. And it belongs to the asteroid belt.

Another note, before examining the other groups, we can calculate the total mass contained in the asteroid belt and divide by two. This would give us a close approximation for the two planets which used to be there (ostensibly, before they collided…or were, somehow, smashed by an incoming immense mass). I propose that the larger planet be named Beverly and the twin named Baby K.

Jupiter has a diameter of about 142,980 km. It has approximately 16 (substantial) moons which I shall not even attempt to calculate for mass because there is a much easier way to calculate which planets should be paired up…as a rough estimate of little consequence.

Because of the angular velocity, or nature of light, it causes a thermal response on each planet in space. This response seems to be absolute in that each planet rotates on its axis in about 24 hrs or less. This allows us to calculate which planets are paired up (or should be).

Jupiter has 16 moons and rotates at just under ten hours. Saturn, diameter 120,540 km, has 18 moons and rotates at just over ten hours. Obvious mates. Ditto for Uranus, 51,120 km, and its 15 moons versus Neptune, 49,530 km, with its 8 moons. Pluto and Charon are similarly matched to each other.

Getting back to Mercury, let us assume, for a brief interlude, that Mercury gets repulsed from the sun. It may be trying to do this now. And that would certainly account for the little abnormality at aphelion.

Anyway, Mercury flies off and is captured by Venus (flies off because of the path torn in space created by the Sun when it entered our system).

Total mass, for moon and planet Venus, is now very nearly identical to Earth and its moon. And space is so structured that each level away from the sun needs two identical masses per "ring" or layer. Hence, little ol' Mars gets shoved back into the asteroid belt.

Meanwhile, back at planet Earth, the Earth pole's reverse and Earth begins to rotate backwards. More than likely, this reversal will occur as a result of the Earth flipping over (so that it is continually spinning as it reverses).

Venus, likewise, reverses its poles and it, too begins spinning opposite to what it is now doing. The solar system is now balanced because Mercury/Venus is the right mass, whereas Mars/Deimos-Phobos is slightly askew.

By the way, the date that Mercury rejoins Venus, as well as the date that the Earth's poles reverse, should be precisely calculated by observing the stutter step of Mercury at Apehelion. Compare length and magnitude. Since Mercury revolves around the sun every 88 days or so, it should be easy enough to do.

Continuing: The base formula for planetary calculations is going to be the same for the atom. This equation is $M(a) + M(b) = M(c) + M(d)$. My computer lacks the function which would allow me to make the small letters proper notations. Just beware that they should be noted accordingly.

M(a) is the mass of the host planet. M(b) is the mass of the moon(s) of M(a). Together, these make the total mass of the planet and it moon(s). And we do the same thing for the other planet-moon(s) paired to it. That is, M(c) is the mass of the second planet, and M(d) is the mass of the second planet's moon(s).

In a very literal, and real, sense, this is the base formula for the Unified Field. While the formula appears to concern itself with mass, as a spatial quality, I am not entirely convinced and feel that more data is required to confirm or deny the spatial aspect.

It is possible that Einstein's famous equation, E=MC(squared), could come in handy for determining the energy to mass ratio. In that case, my Dual Mass equation becomes an energy calculator. My personal belief is that there is a direct correlation between energy and matter which states that it takes X amount of energy to hold X amount of matter together. And this, in turn, will prove that energy and matter are *not* the same.

Bohr's model of the atom is obsolete. Niel's was a remarkably insightful, and intuitive man. All the more so for having made his postulates in an industrial stone-age! And I hope that people never forget the importance of his contribution.

In time, we will come to realize that the nucleus, the proton, and the electron, are all units of mass which are acted upon by units of force. We can now calculate those forces by calculating the masses of each component.

I reiterate that, by looking at what is going on in the solar system, we can determine what is going on at the subatomic level. It is the same force.

Several years ago, I developed a theory about electron pairing. Even if the formula was not so appealing to many, the name is marvelous. It has a nice ring to it. And it is possible.

Going back to the concept of Planetary Pairing, we might well want to reexamine the Dual Aspect theory of light. If we are reluctant to accept the dual atom as being the cause of the disparity, it is easy for us to utilize a dual photon postulate. They are small enough, and fast enough, that two of them could sail through a tiny slit. And when we opened the second slit, the faster second one would fly off, tangentially, and enter the second slot. Piece of cake.

I could go on and on and on. And I may, someday. Right now, I just want to finish this book, get it published and watch the fur fly. It will be fun to watch as hundreds of variants, calculations, and the like, spring forth from my groundwork.

I, for one, am excited to see if anything I may have missed comes to fruition. I predict something along the lines of an atom bomb compared to relativity. Einstein could never have imagined such a thing coming to life in his own era.

I have always considered jet engines to be dinosaurs. As we progress, scientists will realize that space is a crosshatch of lines of force. By studying the cellular levels of the planetary pairs, they should be able to discover how these lines of force operate. Couple that with a more accurate depiction of atomic structure and, well, the sky is no longer the limit.

Cars will make jet propulsion obsolete by tapping into these lines of force. Hover cars? Probably sooner than later.

Isn't it nice to have something to look forward to? Maybe we all will get into the habit of climbing out of bed and reading the paper to see what new discoveries science has made overnight. I know I am.

GASOLINE ENGINES ARE OBSOLETE

An Oregon man claims that gasoline engines are obsolete and have outlived their usefulness. Moreover, according to John Hildreth Atkins, gas powered motors have been headed the way of the dinosaurs for quite some time.

"Forty years ago," Atkins says, "I built a motor out of magnets and some old bicycle parts. It was cool but, even then, I had a sense that it was obsolete, too."

"There are so many energy sources out there, just waiting to be utilized, that it seems absurd to be fighting with OPEC."

"The United States used to be known as the most technological country on the planet. Now, it seems, scientists don't even know how to tie their own shoes."

According to Atkins, physicists have bogged down in a quagmire because they took the path of least resistance. Instead of searching for the truth, they found it easier to build on the work of their predecessors. For instance, it has long been accepted that all atoms carry atoms. That is a fatally erroneous assumption.

Be that as it may, Atkins frequently wonders why these magnet motors are not in every home. They do not pollute. They do not require the consumption of fossil, or any other, fuel. They are cheap to build, require little maintenance, and just plain make sense.

"Corporate greed is the only viable explanation. Magnet motors provide an alternative energy source which would devastate oil companies, as well as commercial power suppliers."

Therein lies the rub. The almighty dollar has replaced good sense. And we are all paying with our health...not to mention our pocketbooks.

GLOBAL WARMING

If the solar system is an accurate depiction of some force of thermodynamics, then we have another factor to consider when examining possible outcomes of global warming. I am, of course, referencing the reversal of the Earth's magnetic poles.

How can I be so sure that global warming is, in reality, a cause of such a catastrophic event? Other than my keen intuition, science will bear it out in several ways. Some of these are reliant upon classical physics. Some are contingent on (new) theoretical physics.

To begin with, we know that heating up an iron magnet will stop its magnetic properties. Heat, as we all know, affects air flow, air flow creates (static) electricity, and electricity creates magnetism. Each is a logical consequence which we easily accept as "fact."

Now we come to the theoretical side of this equation. It is this equation which is suggested by the solar system, itself. More specifically, the motion of the planets.

Looking at the first four planets (Mercury, Venus, Earth, and Mars), we observe that these four planets rotate on their axis at a much slower rate then their four larger brothers (outer four planets). Unfortunately, there are many reasons for these discrepancies.

First, the inner four are moving around the sun at a rate of speed which is much faster than the outer four. Friction might, therefore, be the deciding factor.

Second, the inner four are many times smaller than the outer four. We know that smaller objects (bodies) can, and often do, move faster than their larger constituents.

Third, the surface temperature of the inner four is considerably higher than that of the outer planets. By my estimation, it is this temperature differential which is causing the planets to rotate on their axis.

So? Well, I could be wrong, but it seems to me that thermodynamics plays a pivotal role in a planets movements. If so, then, as our sphere warms up, it should be slowing down.

Picture a ball bearing. If it is dry, it does not function efficiently. Eventually, the ball-bearing will get rusty and stop working. However, we can grease up the ball-bearing and it will function very well again.

Planets, and moons, do not seem to function (rotate on their axis) if they do not have atmospheres (GREASE). Cold surfaces, atmospheres, and size, all seem to contribute to the rate at which a planet moves.

One thing which I am looking at is polar ice. My bet guess is that polar ice is the grease which enables the planets to rotate on their axis more efficiently. If so, then as one of Earth's poles loses its polar ice, may cause the Earth to wobble and tilt. This would explain the wobble and the reversal of the Earth's poles.

Scientists say that our moon is moving away from us. This particular movement would be an expected consequence of Earth's slowing down.

It would be easy for me to take the obvious approach and state that when the Earth slows down enough, the moon will disembark on a journey that is far removed from Earthly experience, and leave it at that. But that isn't my nature. So let's shake it up.

I predict that, as the Earth slows down, the moon is going to get further away from us. Just how far it gets is going to depend on several factors. One of these will be global warming. And that just happens to be the only factor which we have any control over!

The next thing that will occur is that Mercury will get pushed into the sun. Venus will begin to rotate faster when the moon gets within orbital distance of it. To balance out the mass equation,

Earth will capture Mars. Deimos and Phobos will go to Venus.

The resulting destruction will be catastrophic.

In Conclusion

I had actually written the afore-stated articles individual of one another. I had no intention, and certainly no idea, that I was about to unite them all into a common thread. That thread, it would seem, is mainly centered around the concept of speed.

It is common practice to say that there are three states of matter. Gas, liquid, and solid. Nothing difficult about that acknowledgment. But let's re-examine it; shall we?

Each of those three states is in reference to three-dimensional matter. With a little bit of a nudge, we can easily see where gas can be classified as an attempt to transmute from three dimensions to two dimensions. In other words, the closer we come to eliminating one of the dimensions, the greater we increase our speed.

As a side-note, at the time that I write this, I am indecisive as to which end of the dimensional spectrum radioactive material would be placed. My own inclination is to put it at the bottom (slow) end. Perhaps it is so tightly packed that it is collapsing in on itself. But we won't worry about radioactive material at this juncture.

I realize that the natural progression is from slow to fast, from solid to gas, and there is sound reasoning behind this transformation. In its simplest form, we are adding energy. As we add energy, by definition, we are increasing activity and all of that energy (pardon the pun) has to go somewhere.

My earlier declaration was that magnetism is slow (for two-dimensional existence. If gas is also at the cusp, then it stands to reason that there might be some symbiotic relationship between magnetism and gas. Gas tends to be hot and magnetism tends to be cold. Ignoring that possibility, we are still left with an approximate speed limit for that border.

I am sure that many of you have already postulated that if Matter has three states, then, perhaps, energy must, too. And, if you are particularly clever, you will have surmised that each dimension has three (or more) states. And, if you are really wise, you will come to the conclusion that residents of the first, and second, dimensions must be energy...or, at least, more energy than matter.

As speed, velocity, or acceleration increase with the decrease in the number of dimensions, another factor arises. Matter has, suddenly, become energy. And the faster that a particle moves, the less energy it takes to move it---until you reach its saturation point. Let's hold that thought for a minute.

We all know that it takes a computed infinite energy to move matter at the speed of light. The reality is, you cannot move matter at the speed of light. Matter will convert to energy long before then. It will also shed one, or more, dimensions.

Energy, or the coefficient of energy, is just the opposite. Why? Because energy, in its purest form, travels at infinite, or near infinite, speed. You cannot pile an infinite on top of an infinite.

Now this creates an unusual situation in that it takes energy in order to slow down energy. This energy exchanges for matter rather nicely. That is to say, a very small piece of matter can stop a very large piece of energy. So? What am I getting at?

The point that I am so miserably trying to put across is simply that as light slows down, it increases in mass/force. We can actually derive at that conclusion from another direction. Take the case of our sun, for example. It houses a tremendous amount of energy which it uses to expel photons. Since energy is conserved, whatever force expended by the sun is going to be the same force required to slow the photon.

From the atom, we now return to the cosmos. I bring you back to this point because I have a little theory which I am sure will interest others. I theorize that there is an invisible barrier somewhere between Mars and Jupiter. This barrier only allows bodies under 13,000km to pass through unscathed. Larger bodies are ripped apart.

This invisible barrier is, most likely, the outermost shell of an electron. The barrier is more of a line of protection than a means of containing whatever happens to be inside of its protected region.

I predict that the barrier is closer to Jupiter. In fact, I believe that it is so close that it may even affect some of Jupiter's moons (particularly the larger ones). I leave that to scientists to discover.

I used to tell people that I could provide them with the answer to their question---if they could explain their question to me. This was my way of bypassing the education that I knew I was lacking. For instance, I had never heard of Brownian Movement and, therefore, had little hope of being able to explain it.

I have since discovered that Brownian Movement is a jittery dance that some solids seem to exhibit while suspended in a liquid. It has become clear to me and, from there, I have derived at a unilateral relationship between energy and matter, and not-so-steady state.

My point is not, despite contrary opinion, to brag about some imaginary intellectual prowess which I may possess. My point is to impress upon my readers a rule which Physicists have long since abandoned. That is: Make sure that you understand what it is you think you are seeing.

Sir Isaac Newton is said to have observed an apple falling to the ground. From that event, he speculated that the moon attracted the earth. And we all know the result of that observance.

Newton went on to postulate that gravity was a force inherent in all things. He defined this force as being an attraction between two bodies. The bigger the bodies, and the closer they got to each other, the greater the force of attraction. Just why the moon did not plummet to the Earth is anybody's guess.

Newton's observations satisfied almost all of the tenants of accepted science. Namely, it was a 'fact' that could be observed by practically everyone who desired to do so. And, so long as we were only discussing large bodies in space, it seemed to be predictable.

All seceding scientists adhered to Newton's theories because they had appeared to hold up. And the parts which did not hold up were, for the most part, ignored or shrugged off. Instead of looking for an alternate explanation, it was easier to look for the 'proverbial' exception to the rule. In the final analysis, that turned out to be most unscientific.

Not to detract from the very important work of talented men like Lorentz, Poincare, Planck, Hertz, etc., etc., but it seems to me absolutely amazing that nobody sought to challenge Newton. Had they done so, how very different science would be today.

My theories are not so much the result of extraordinary intelligence as they are the product of a recalcitrant mind. Of course, it didn't hurt the cause any that I was spared the 'higher education' that my predecessors reveled in. Who knows what damages a college education may have done to me?

I have found one thing that is peculiar to all living beings. No matter how much we learn, there is always going to be more to learn. Take, for instance, the ant. He goes out in search of food. When he discovers it, he calls in his buddies to harvest it. Then, he continues with his quest.

I find that particularly reassuring. How redundant the world would be if we knew everything there was to know. Every quest is like shaking a Christmas gift. We guess at what is inside and, in that precise moment while we are untying the ribbon, it doesn't really matter what is inside the box. We have the ability to imagine anything we want.

Here's hoping that you never stop shaking the boxes. And, if you do stop, let it be because you found whatever it was you were after!

johnhildrethatkins@yahoo.com

Spiel:

Written in plain and concise language which everyone can understand. Very satisfying.

www.ingramcontent.com/pod-product-compliance
Lightning Source LLC
Chambersburg PA
CBHW020656220526
45464CB00001B/459